SpringerBriefs in Applied Sciences
and Technology

Continuum Mechanics

Series Editors

Holm Altenbach, Institut für Mechanik, Lehrstuhl für Technische Mechanik,
Otto von Guericke University Magdeburg, Magdeburg, Sachsen-Anhalt, Germany

Andreas Öchsner, Faculty of Mechanical Engineering, Esslingen University of
Applied Sciences, Esslingen am Neckar, Germany

These SpringerBriefs publish concise summaries of cutting-edge research and practical applications on any subject of Continuum Mechanics and Generalized Continua, including the theory of elasticity, heat conduction, thermodynamics, electromagnetic continua, as well as applied mathematics.

SpringerBriefs in Continuum Mechanics are devoted to the publication of fundamentals and applications, presenting concise summaries of cutting-edge research and practical applications across a wide spectrum of fields. Featuring compact volumes of 50 to 125 pages, the series covers a range of content from professional to academic.

Sergey Korobeynikov · Alexey Larichkin

Objective Algorithms for Integrating Hypoelastic Constitutive Relations Based on Corotational Stress Rates

Springer

Sergey Korobeynikov
Lavrentyev Institute of Hydrodynamics
Novosibirsk, Russia

Alexey Larichkin
Lavrentyev Institute of Hydrodynamics
Novosibirsk, Russia

ISSN 2191-530X   ISSN 2191-5318   (electronic)
SpringerBriefs in Applied Sciences and Technology
ISSN 2625-1329   ISSN 2625-1337   (electronic)
SpringerBriefs in Continuum Mechanics
ISBN 978-3-031-29631-4   ISBN 978-3-031-29632-1   (eBook)
https://doi.org/10.1007/978-3-031-29632-1

This Springer imprint is published by the registered company Springer Nature Switzerland AG
The registered company address is: Gewerbestrasse 11, 6330 Cham, Switzerland

Preface

This book provides readers with a deep understanding of the use of objective algorithms for integration of constitutive relations (CRs) for Hooke-like hypoelasticity based on the use of corotational stress rates. The purpose of objective algorithms is to perform the step-by-step integration of CRs using fairly large time steps that provide high accuracy of this integration in combination with the exact reproduction of superimposed rigid body motions. Since Hooke-like hypoelasticity is included as a component in CRs for elastic-inelastic materials (e.g., in CRs for elastic-plastic materials), the scope of these algorithms is not limited to hypoelastic materials, but extends to many other materials subjected to large deformations. The authors performed a comparative analysis of the performance of the most currently available objective algorithms, provided some recommendations for improving the existing formulations of these algorithms, and presented new formulations of the so-called absolutely objective algorithms. The proposed book will be useful for beginner researchers in the development of economical methods for integrating elastic-inelastic CRs, as well as for experienced researchers, by providing a compact overview of existing objective algorithms and new formulations of these algorithms. The book will also be useful for developers of computer codes for implementing objective algorithms in FE systems. In addition, this book will also be useful for users of commercial FE codes, since often these codes are so-called black boxes and this book shows how to test accuracy of the algorithms of these codes for integrating elastic-inelastic CRs in modeling large rotations superimposed on the uniform deformation of any sample.

Novosibirsk, Russia

Sergey Korobeynikov
Alexey Larichkin

Acknowledgements

The authors thank Prof. Otto T. Bruhns (Ruhr University, Bochum, Germany) for a careful reading of the manuscript of this book and for a number of useful comments that have significantly improved the final version of the manuscript. The authors also thank Prof. Holm Altenbach (Otto von Guericke University, Magdeburg, Germany), who, as co-editor of the SpringerBriefs in Continuum Mechanics series, helped the authors find an excellent platform to present their research results published in this book. In addition, we are also grateful to Dr. Tatyana A. Rotanova (Novosibirsk State University, Russia) for collaboration in the implementation of hypoelasticity models in the homemade FE code.

The support from the Russian Science Foundation (Grant No. 22-29-00485, https://rscf.ru/project/22-29-00485/) is gratefully acknowledged.

Novosibirsk, Russia   Sergey Korobeynikov
January 2023   Alexey Larichkin

Contents

Acronyms

CRs   Constitutive relations
ETs   Euclidean transformations
FE   Finite element
FEM   Finite element method
H–W   Hughes and Winget
IETs   Incremental Euclidean transformations
R–A   Rubinstein and Atluri
S–H   Simo and Hughes
SRBMs   Superimposed rigid body motions

Chapter 1
Introduction

Abstract Objective algorithms for integrating constitutive relations for inelastic material models have been developed for over 40 years; therefore, the state of the art of these studies is presented in Sect. 1.1. In Sect. 1.2 we formulate the objectives of this study and give a brief overview of the development of the considered algorithms and new (previously unpublished) objective algorithms developed by us for integrating hypoelastic constitutive relations based on corotational stress rates.

1.1 State of the Art

In step-by-step procedures for solving quasi-static and implicit dynamic equations with an iterative refinement of the solution by the Newton–Raphson method, the time step is mainly limited by the accuracy of the numerical integration of these equations. Therefore, the efforts of most developers of algorithms for the numerical solution of these equations have focused on developing numerical integration schemes for solving problems of deformation of solids and structures with the largest possible time step and a minimal loss of accuracy. This is especially true for the integration of constitutive relations (CRs) for inelastic material models at finite element integration points due to the large amount of computations required for the integration.

At the dawn of the development of finite element (FE) codes (1970s) based on implicit formulations of the finite element method (FEM) for solving nonlinear problems of solid mechanics, the simplest Euler scheme with first-order accuracy in time was mainly used for explicit integration of CRs for inelastic material models (see, e.g., [1]). At the same time, in FE codes implementing explicit formulations of the FEM (as in implementations of the molecular dynamics method, see, e.g., [2]), the midpoint algorithm with second-order accuracy in time was used for explicit integration of CRs for inelastic material models. However, when using large time steps in the midpoint algorithm for integrating CRs for inelastic material models in FE codes based on implicit formulations of the FEM, there is a problem of accurately determining stresses under superposed rigid body motions (SRBMs) (i.e., under large rotations of material fibers superimposed on both small and large deformations of solids, cf., [3–11]).

© The Author(s), under exclusive license to Springer Nature Switzerland AG 2023   1
S. Korobeynikov and A. Larichkin, *Objective Algorithms for Integrating Hypoelastic Constitutive Relations Based on Corotational Stress Rates*, SpringerBriefs in Continuum Mechanics, https://doi.org/10.1007/978-3-031-29632-1_1

The CRs for hypoelasticity occupy an intermediate position between the rate formulations of CRs for Cauchy/Green elasticity and elasto-plasticity. Since CRs for hypoelasticity are included as an integral part in CRs for additive elasto-plasticity (see, e.g., [12, 13]), researchers continue to study the properties of these models and numerical integration algorithms for CRs for these models that are already implemented in existing commercial FE codes (cf., [14–16]) as well as continue to develop theoretical background for the formulation of hypoelastic CRs (see, e.g., [17–22]) and algorithms for their numerical integration (see, e.g., [9–11]).

In this book, we analyze existing algorithms and propose advanced algorithms for the numerical integration of CRs for hypoelastic material models. The family of CRs for hypoelastic models can be divided into two subfamilies: the subfamily of CRs for classical hypoelastic models that was introduced by Truesdell in the mid-1950s (cf., [23, 24], see also [25]) and which does not require the introduction of the reference configuration of a deformable body and the subfamily of CRs for generalized hypoelastic models (see, e.g., [26, 27]), for which the introduction of the reference configuration of this body is allowed.[1] In this paper, we present algorithms for the numerical integration of CRs both for the classical hypoelastic models based on the Zaremba–Jaumann and Hill (corotational) stress rates (below, these material models are referred to as the *C-models* for brevity) and for models of generalized hypoelasticity based on corotational stress rates associated with spin tensors from the family of continuous material spin tensors (cf., [29–31]) (below, these material models are referred to as *G-models*). The family of G-models contains an infinite number of models based on an infinite number of spin tensors associated with these models. We will study in detail algorithms for integrating CRs for two material models from this family, namely, the material models based on the Green–Naghdi and logarithmic stress rates, which are widely used in the continuum mechanics equations.

Note that hypoelastic material models based on corotational stress rates occupy a special place in the family of hypoelastic models, since, on the one hand, Prager's principle (cf., [32, 33], see also [12, 13]) allows the use of only these hypoelastic material models in CRs for elasto-plasticity and, on the other hand, Fiala [18–22] has established that of the objective stress rates, only the corotational Zaremba–Jaumann stress rates have a geometrical meaning.

Apparently, the work by Hughes and Winget (cf., [34]) was the first attempt to combine the midpoint algorithm with a reasonably accurate integration of CRs for the classical hypoelastic model based on the Zaremba–Jaumann stress rate for large increments of strain and rotation of material fibers of deformable solids. Hughes and Winget defined their algorithm as an *incrementally objective* algorithm. The essence of this algorithm is that in the absence of rotations, the stresses due to large strains are integrated over time with second-order accuracy, and in the absence of strains, the rotations of material fibers are also integrated fairly accurately and spurious increments of stresses due to inaccurate determination of rotation increments do not

---

[1] See the discussion in Sect. 1 in [28].

occur. The development of the Hughes–Winget algorithm (the *H–W algorithm*) was continued in [17, 35–44].

However, already in the early 1980s, doubts arose that the H–W algorithm would fairly accurately simulate stresses under the simultaneous action of strains and SRBMs (c.f., [7, 8]). Indeed, Rashid [6] have found that for strains with SRBMs, the accuracy of the H–W algorithm in determining stresses is much lower than that for strains without SRBMs. It has been proposed [6–8] to call the H–W algorithm a *weak incrementally objective* algorithm and algorithms that exactly reproduce SRBMs *strong incrementally objective* algorithms. Strong incrementally objective algorithms for integrating CRs for hypoelastic models based on corotational and non-corotational stress rates have been developed [3–6, 9–11, 35, 45–53]. However, in these papers, algorithms for integrating CRs only for hypoelasticity models based on the Zaremba–Jaumann stress rate are presented, but similar algorithms for integrating CRs for the entire family of G-models are not proposed.

Note that hypoelastic CRs should satisfy the Lagrangian or Eulerian objectivity conditions (see, e.g., [28, 54]). However, strong incrementally objective algorithms can only be used to integrate CRs for hypoelastic models based on the Eulerian formulation. To provide a solid theoretical background for the further development of such algorithms, in this book we introduce definitions of the *incremental objectivity* (Lagrangian, Eulerian, Eulerian–Lagrangian, and Lagrangian–Eulerian) of tensors, which are then used to generalize strong incrementally objective algorithms for integrating CRs for G-models based on the Eulerian formulation.

Algorithms for integrating CRs for hypoelastic models based on the Green–Naghdi stress rate in both the Lagrangian and Eulerian formulations are developed in [17, 35, 37, 38, 40, 55]. In particular, in these papers, the approximate incremental strain tensor is determined (as in the weak incrementally objective algorithm) using the classical midpoint second-order accurate rule, but the incremental rotation tensor is determined without any approximations, directly from the polar decomposition of the deformation gradient tensor (in [35], this algorithm for integrating CRs is called the GN1 algorithm). As a result, although this algorithm is improved compared to the standard weak incrementally objective algorithms, but, like the standard weak incrementally objective algorithms, it also does not accurately reproduce SRBMs (see the plots in Fig. 12 in [35] obtained using the GN1 algorithm in a computer simulation of simple extension with superimposed rotation for a square hypoelastic sample with CRs for the model based on the Green–Naghdi stress rate). In addition, the GN2 algorithm (alternative to the GN1 algorithm) for integrating CRs for this material model was developed in [35]. In this algorithm, the approximate incremental strain tensor is determined using the generalized midpoint first-order accurate algorithm (as in the strong incrementally objective algorithm), but the incremental rotation tensor is determined, as in the GN1 algorithm, without any approximations, directly from the polar decomposition of the deformation gradient tensor. As a result, the GN2 algorithm is strong incrementally objective, and it already exactly reproduces SRBMs (see the plots in Fig. 12 in [35] obtained using the GN2 algorithm in a computer simulation of simple extension with a superimposed rotation for a square hypoelastic sample with CRs for the model based on the Green–Naghdi stress rate).

1.2   Book Writing Goals

In this book, we present a new family of algorithms for integrating CRs for hypoe-lastic models based on corotational stress rates associated with spin tensors from the family of continuous material spin tensors (cf., [29–31])—a family of *absolutely objective algorithms*. The construction of this family of algorithms is based on the idea that approximations of incremental strain and rotation tensors should have the same type of absolute (not incremental) objectivity as their continuous counterparts. Algorithms of the new family exactly reproduce SRBMs and in this sense are com-parable to strong incrementally objective algorithms. These algorithms, like strong incrementally objective algorithms, can be used to integrate CRs for G- and C-models of hypoelasticity. Therefore, an important objective of this study is to establish fam-ilies of material models for which it is preferable to use either strong incrementally objective or absolutely objective algorithms.

The present study has the following main objectives:

1. to determine Lagrangian, Eulerian, Eulerian–Lagrangian, and Lagrangian–Eulerian incremental objectivities for second-order Eulerian tensors;
2. to show that the second-order accurate midpoint rule for integrating CRs for hypoelastic material models can be used to formulate only weak (not strong) incrementally objective algorithms;
3. to show that all previously published strong incrementally objective algorithms for integrating CRs for the hypoelastic model based on the Zaremba–Jaumann stress rate can be obtained using the generalized midpoint rule (cf., [56]) and also show that any strong incrementally objective algorithms for integrating stresses can only be first-order accurate in time;
4. to generalize strong incrementally objective algorithms for integrating CRs for any G-models that reduce to the previously published algorithms for integrating CRs for the hypoelastic model based on the Zaremba–Jaumann stress rate;
5. to develop new absolutely objective algorithms for integrating CRs for hypoelastic models;
6. to conduct a comparative analysis of the performance of the newly developed algo-rithms that exactly reproduce SRBMs in the solution of the simple extension and simple shear problems for hypoelastic models based on the Zaremba–Jaumann, Green–Naghdi, logarithmic, and Hill stress rates and determine which of the gen-eralized and newly developed algorithms are most suitable for integrating CRs for these hypoelastic models.

The main results of this study are as follows:

(1) new forms of the approximate incremental strain tensor—the *Mooney incremental strain tensors*—were introduced which improved the performance of strong incre-mentally objective algorithms; (2) a generalization was made of strong incrementally objective algorithms for integrating CRs for hypoelastic material models based on corotational stress rates associated with spin tensors dependent on the choice of the reference configuration; (3) new absolutely objective algorithms were developed

which compete with strong incrementally objective algorithms; (4) the results of computer simulations of simple extension and simple shear show that strong incrementally objective are preferred to absolutely objective algorithms for integrating CRs for hypoelastic models based on kinematic variables independent of the choice of the reference configuration. At the same time, absolutely objective algorithms are preferred for integrating CRs for hypoelastic models based on kinematic variables dependent on the choice of the reference configuration.

Flowcharts of all algorithms considered in this book are given in Appendix A.

References

1. K.J. Bathe, *Finite Element Procedures* (Prentice Hall, Upper Saddle River, New Jersey, 1996)
2. M.J. Buehler, *Atomistic Modeling of Materials Failure* (Springer, New York, 2008)
3. M. Hollenstein, M. Jabareen, M.B. Rubin, Comput. Mech. **52**, 649 (2013). https://doi.org/10.1007/s00466-013-0838-7
4. M. Jabareen, Int. J. Eng. Sci. **96**, 46 (2015). https://doi.org/10.1016/j.ijengsci.2015.07.001
5. M. Kroon, M.B. Rubin, Finite Elem. Anal. Des. **177**, 103422 (2020). https://doi.org/10.1016/j.finel.2020.103422
6. M.M. Rashid, Int. J. Numer. Methods Eng. **36**, 3937 (1993). https://doi.org/10.1002/nme.1620362302
7. K.W. Reed, S.N. Atluri, Comput. Methods Appl. Mech. Eng. **39**, 245 (1983). https://doi.org/10.1016/0045-7825(83)90094-4
8. K.W. Reed, S.N. Atluri, Int. J. Plast. **1**, 63 (1985). https://doi.org/10.1016/0749-6419(85)90014-2
9. M.B. Rubin, O. Papes, J. Mech. Mater. Struct. **6**, 529 (2011). https://doi.org/10.2140/jomms.2011.6.529
10. M.B. Rubin, Finite Elem. Anal. Des. **175**, 103409 (2020). https://doi.org/10.1016/j.finel.2020.103409
11. M.B. Rubin, *Continuum Mechanics with Eulerian Formulations of Constitutive Equations* (Springer, Cham, 2021)
12. O.T. Bruhns, Z. Angew. Math. Mech. **94**, 187 (2014). https://doi.org/10.1002/zamm.201300243
13. O.T. Bruhns, Acta. Mech. Sin. **36**, 472 (2020). https://doi.org/10.1007/s10409-020-00926-7
14. L. Gambirasio, G. Chiantoni, E. Rizzi, Arch. Computat. Methods Eng. **23**, 39 (2016). https://doi.org/10.1007/s11831-014-9130-z
15. W. Ji, A.M. Waas, Z.P. Bažant, J. Appl. Mech. **80**, 041024 (2013). https://doi.org/10.1115/1.4007828
16. N. Nguyen, A. Waas, Z. Angew. Math. Phys. **67**, 35 (2016). https://doi.org/10.1007/s00033-016-0623-5
17. D. Aubram, *Notes on Rate Equations in Nonlinear Continuum Mechanics* (2017). https://arxiv.org/abs/1709.10048
18. Z. Fiala, Ann. Phys. (N. Y.) **326**, 1983 (2011). https://doi.org/10.1016/j.aop.2011.02.010
19. Z. Fiala, Acta Mech. **226**, 17 (2015). https://doi.org/10.1007/s00707-014-1162-9
20. Z. Fiala, Int. J. Non Linear Mech. **81**, 230 (2016). https://doi.org/10.1016/j.ijnonlinmec.2016.01.019
21. Z. Fiala, in *Emerging Concepts in Evolution Equations*, ed. by C. Murphy (Nova Science, New York, 2017), pp. 1–30
22. Z. Fiala, Z. Angew. Math. Phys. **71**, 4 (2020). https://doi.org/10.1007/s00033-019-1227-7
23. C. Truesdell, Commun. Pure Appl. Math. **8**, 123 (1955). https://doi.org/10.1002/cpa.3160080109

24. C. Truesdell, J. Ration. Mech. Anal. **4**, 83 (1955)
25. C. Truesdell, W. Noll, in *Handbuch der Physik, Vol. III/3*, ed. by S. Flügge (Springer, Berlin, 1965), pp. 1–602
26. A.E. Green, B.C. McInnis, P. Roy. Soc. Edinb. A **67**, 220 (1967). https://doi.org/10.1017/S0080454100008074
27. R. Hill, J. Mech. Phys. Solids **7**, 209 (1959). https://doi.org/10.1016/0022-5096(59)90007-92
28. S.N. Korobeynikov, Arch. Appl. Mech. **90**, 313 (2020). https://doi.org/10.1007/s00419-019-01611-3
29. S.N. Korobeynikov, Acta Mech. **216**, 301 (2011). https://doi.org/10.1007/s00707-010-0369-7
30. H. Xiao, O.T. Bruhns, A. Meyers, Int. J. Solids Struct. **35**, 4001 (1998). https://doi.org/10.1016/S0020-7683(97)00267-9
31. H. Xiao, O.T. Bruhns, A. Meyers, J. Elast. **52**, 1 (1998). https://doi.org/10.1023/A:1007570827614
32. W. Prager, Quart. Appl. Math. **18**, 403 (1960)
33. W. Prager, *Introduction to Mechanics of Continua* (Dover Publications, Mineola, 2004)
34. T.J.R. Hughes, J. Winget, Int. J. Numer. Methods Eng. **15**, 1862 (1980). https://doi.org/10.1002/nme.1620151210
35. J. Dabounou, J. Eng. Mech. **142**, 04016056 (2016). https://doi.org/10.1061/(ASCE)EM.1943-7889.0001112
36. J. Fish, K. Shek, Int. J. Numer. Methods Eng. **44**, 839 (1999). https://doi.org/10.1002/(SICI)1097-0207(19990228)44:6<839::AID-NME533>3.0.CO;2-C
37. B.E. Healy, R.H. Dodds Jr., Comput. Mech. **9**, 95 (1992). https://doi.org/10.1007/BF00370065
38. T.J.R. Hughes, in *Theoretical Foundation for Large-scale Computations for Nonlinear Material Behavior*, ed. by S. Nemat-Nasser et al. (Martinus Nijhoff Publishers, Dordrecht, 1984), pp. 29–63
39. S.W. Key, R.D. Krieg, Comput. Methods Appl. Mech. Eng. **33**, 439 (1982). https://doi.org/10.1016/0045-7825(82)90118-9
40. S.H. Lo, Int. J. Numer. Methods Eng. **26**, 121 (1988). https://doi.org/10.1002/nme.1620260109
41. P. Longère, Mech. Res. Commun. **95**, 61 (2019). https://doi.org/10.1016/j.mechrescom.2018.12.001
42. M. Nazem, J.P. Carter, D. Sheng, S.W. Sloan, Finite Elem. Anal. Des. **45**, 934 (2009). https://doi.org/10.1016/j.finel.2009.09.006
43. P.M. Pinsky, M. Ortiz, K.S. Pister, Comput. Methods Appl. Mech. Eng. **40**, 137 (1983). https://doi.org/10.1016/0045-7825(83)90087-7
44. R. Rubinstein, S.N. Atluri, Comput. Methods Appl. Mech. Eng. **36**, 277 (1983). https://doi.org/10.1016/0045-7825(83)90125-1
45. M.S. Gadala, J. Wang, Finite Elem. Anal. Des. **35**, 379 (2000). https://doi.org/10.1016/S0168-874X(00)00003-2
46. M. Kleiber, P. Kowalczyk, *Introduction to Nonlinear Thermomechanics of Solids* (Springer, Switzerland, 2016)
47. A.M. Lush, G. Weber, L. Anand, Int. J. Plast. **5**, 521 (1989). https://doi.org/10.1016/0749-6419(89)90012-0
48. J.C. Nagtegaal, Comput. Methods Appl. Mech. Eng. **33**, 469 (1982). https://doi.org/10.1016/0045-7825(82)90120-7
49. J.C. Nagtegaal, F.E. Veldpaus, in *Numerical Methods in Industrial Forming Processes*, ed. by J. Pittman (Wiley, Swansea, 1984), pp. 351–371
50. A. Rodriguez-Ferran, P. Pegon, A. Huerta, Int. J. Numer. Methods Eng. **40**, 4363 (1997). https://doi.org/10.1002/(SICI)1097-0207(19971215)40:23<4363::AID-NME263>3.0.CO;2-Z
51. A. Rodriguez-Ferran, A. Huerta, J. Eng. Mech. **124**, 939 (1998). https://doi.org/10.1061/(ASCE)0733-9399(1998)124:9(939)
52. G.G. Weber, A.M. Lush, A. Zavaliangos, L. Anand, Int. J. Plast. **6**, 701 (1990). https://doi.org/10.1016/0749-6419(90)90040-L

53. X. Zhou, K.K. Tamma, Finite Elem. Anal. Des. **39**, 783 (2003). https://doi.org/10.1016/S0168-874X(03)00059-3
54. S.N. Korobeynikov, J. Elast. **143**, 147 (2021). https://doi.org/10.1007/s10659-020-09808-2
55. D.P. Flanagan, L.M. Taylor, Comput. Methods Appl. Mech. Eng. **62**, 305 (1987). https://doi.org/10.1016/0045-7825(87)90065-X
56. J.C. Simo, T.J.R. Hughes, *Computational Inelasticity* (Springer, N.Y., 1998)

Chapter 2
Preliminaries

Abstract In this section, we present the background from continuum mechanics required for the development of objective algorithms to integrate CRs for hypoelasticity models. In Sect. 2.1, we consider the objective (Lagrangian and Eulerian) strain tensors known in the literature. In Sect. 2.2, we present objective strain rates and formulate some propositions that will be used to establish the laws of multiplicative decomposition of rotation tensors associated with spin tensors by additive decompositions of these spin tensors. These propositions will allow us to further generalize strong incrementally objective algorithms and formulate new absolutely objective algorithms for integrating CRs for hypoelastic material models based on corotational stress rates associated with spin tensors from the family of continuous material spin tensors. In Sect. 2.3, we present formulations of CRs for hypoelastic material models in a form convenient for their further integration. In particular, corotational stress rates are represented in the form of Lie time derivatives. In addition, we divide the CRs into components that are calculated exactly and those that require numerical integration. In the same section, CRs are formulated for hypoelastic material models based on the Zaremba–Jaumann, Green–Naghdi, logarithmic, and Hill stress rates.

2.1 Local Body Deformations and Basic Kinematics

Consider the motion of a body $\mathfrak{B}$ in a three-dimensional Euclidean point space, and let $\mathbf{X}$ and $\mathbf{x}$ be the position vectors of some particle $P \in \mathfrak{B}$ in the *reference* (fixed at time t_0) and *current* (moving at time t) configurations, respectively. Let $\mathbf{F} \equiv \mathrm{Grad}\, \mathbf{x} = \partial \mathbf{x}/\partial \mathbf{X} \in \mathcal{T}^{21}$ ($J \equiv \det \mathbf{F} > 0$) be the *deformation gradient*. The tensor $\mathbf{F}$ can be uniquely represented as (see, e.g., [1–4])[2]

$$\mathbf{F} = \mathbf{R} \cdot \mathbf{U} = \mathbf{V} \cdot \mathbf{R} \quad (\mathbf{U}^T = \mathbf{U},\ \mathbf{V}^T = \mathbf{V},\ \mathbf{R} \cdot \mathbf{R}^T = \mathbf{I},\ \det \mathbf{R} = 1), \qquad (2.1)$$

---

[1] Hereinafter, $\mathcal{T}^2$ denotes the set of all second-order tensors.

[2] Hereinafter, the superscript T denotes the transposition of a tensor.

© The Author(s), under exclusive license to Springer Nature Switzerland AG 2023
S. Korobeynikov and A. Larichkin, *Objective Algorithms for Integrating Hypoelastic Constitutive Relations Based on Corotational Stress Rates*, SpringerBriefs in Continuum Mechanics, https://doi.org/10.1007/978-3-031-29632-1_2

where $\mathbf{U}$, $\mathbf{V} \in \mathcal{T}_{\mathrm{sym}}^{2}$[3] are the *right* (Lagrangian) and *left* (Eulerian) positive definite *stretch tensors*, respectively, and $\mathbf{R} \in \mathcal{T}_{\mathrm{orth}}^{2+}$ is the *rotation tensor*.[4]

The *spectral representations* (see, e.g., [1, 2, 5–7]) of the tensors $\mathbf{U}$ and $\mathbf{V}$ have the form

$$\mathbf{U} = \sum_{i=1}^{m} \lambda_i \mathbf{U}_i, \quad \mathbf{V} = \sum_{i=1}^{m} \lambda_i \mathbf{V}_i \quad (0 < \lambda_i < \infty). \tag{2.2}$$

Hereinafter, λ_i are all different m[5] eigenvalues (principal stretches) of the tensors $\mathbf{U}$ and $\mathbf{V}$, and $\mathbf{U}_i$ and $\mathbf{V}_i$ are the subordinate *eigenprojections* of these tensors ($i = 1, \ldots, m$).

Remark 2.1 For applications, it is possible to determine (see, e.g., [8]) the *right Cauchy–Green deformation tensor* $\mathbf{C} \equiv \mathbf{F}^T \cdot \mathbf{F} = \mathbf{U}^2$ (using Lagrangian variables) or the *left Piola deformation tensor* $\mathbf{b} \equiv \mathbf{F}^{-T} \cdot \mathbf{F}^{-1} = \mathbf{V}^{-2}$ (using Eulerian variables), and then obtain spectral representations of the tensors $\mathbf{C}$ or $\mathbf{b}$

$$\mathbf{C} = \sum_{i=1}^{m} \mu_i \mathbf{C}_i, \quad \text{or} \quad \mathbf{b} = \sum_{i=1}^{m} \kappa_i \mathbf{b}_i, \tag{2.3}$$

and determine the eigenpairs $(\lambda_i, \mathbf{U}_i)$ or $(\lambda_i, \mathbf{V}_i)$ $(i = 1, \ldots, m)$ from the equalities (see, e.g., [9], Eq. (89))

$$\lambda_i = \sqrt{\mu_i}, \ \mathbf{U}_i = \mathbf{C}_i, \quad \text{or} \quad \lambda_i = 1/\sqrt{\kappa_i}, \ \mathbf{V}_i = \mathbf{b}_i. \tag{2.4}$$

Algorithms for determining the tensors $\mathbf{U}$, $\mathbf{V}$, and $\mathbf{R}$ based on spectral representations of the tensors $\mathbf{C}$ and $\mathbf{b}$ and finding their principal invariants (for 2D analysis) are presented in Appendix B (see also [10–13]).

Consider the *Euclidean transformations* (ETs) (cf., [3, 14–16])[6]

$$\mathbf{x}^*(\mathbf{X}^*, t^*) \equiv \mathbf{Q}(t) \cdot \mathbf{x}(\mathbf{X}, t) + \mathbf{c}(t), \quad \mathbf{X}^* \equiv \mathbf{Q}_0 \cdot \mathbf{X} + \mathbf{c}_0, \quad t^* = t + a, \tag{2.5}$$

where $\mathbf{Q}(t) \in \mathcal{T}_{\mathrm{orth}}^{2+}$ is an arbitrary tensor, $\mathbf{c}(t)$ is an arbitrary vector, and $a \in \mathbb{R}$ is an arbitrary variable ($\mathbf{Q}_0 \equiv \mathbf{Q}(t_0)$ and $\mathbf{c}_0 \equiv \mathbf{c}(t_0)$). We now give definitions of objective tensors following [3, 15, 17, 18].

---

[3] Hereinafter, $\mathcal{T}_{\mathrm{sym}}^{2} \subset \mathcal{T}^{2}$ denotes the set of all symmetric second-order tensors.

[4] Hereinafter, the subset $\mathcal{T}_{\mathrm{orth}}^{2+} \subset \mathcal{T}^{2}$ denotes the set of all proper orthogonal second-order tensors (i.e., the tensors $\mathbf{\Psi}$ such that $\mathbf{\Psi} \cdot \mathbf{\Psi}^T = \mathbf{I}$ and $\det \mathbf{\Psi} = 1$).

[5] The number m $(1 \leq m \leq 3)$ will be called the *eigenindex*.

[6] In Eq. (2.5), it is usually assumed that $\mathbf{Q}_0 = \mathbf{I}$ (see, e.g., [4, 17]). It is, however, more logical to take into account the possibility of variation of the reference configuration of the form $(2.5)_2$. Ogden [3] (Eq. (2.1.20)) only gives an equation of the form $(2.5)_2$ but does not use this equation in the subsequent derivation of the law of variation of Lagrangean tensors under Euclidean transformations.

Definition 2.1 A tensor $\mathbf{H} \in \mathcal{T}^2$ is called an *objective tensor* if, under ETs (2.5), it changes as follows[7]:

$$
\begin{aligned}
\mathbf{H}^*(P, t) &= \mathbf{Q}(t) \cdot \mathbf{H}(P, t) \cdot \mathbf{Q}^T(t) & Eulerian, \qquad\qquad (2.6)\\
\mathbf{H}^*(P, t) &= \mathbf{Q}_0 \cdot \mathbf{H}(P, t) \cdot \mathbf{Q}_0^T & Lagrangian,\\
\mathbf{H}^*(P, t) &= \mathbf{Q}(t) \cdot \mathbf{H}(P, t) \cdot \mathbf{Q}_0^T & Eulerian-Lagrangian,\\
\mathbf{H}^*(P, t) &= \mathbf{Q}_0 \cdot \mathbf{H}(P, t) \cdot \mathbf{Q}^T(t) & Lagrangian-Eulerian.
\end{aligned}
$$

The simple [8] isotropic tensor functions

$$
\mathbf{E} \equiv \mathbf{f}(\mathbf{U}) = \sum_{i=1}^{m} f(\lambda_i) \mathbf{U}_i, \quad \mathbf{e} \equiv \mathbf{f}(\mathbf{V}) = \sum_{i=1}^{m} f(\lambda_i) \mathbf{V}_i,
$$

$$
f(\lambda) \in \mathbb{R}: \ f(\lambda) \in C^2, \quad f'(\lambda) > 0 \text{ as } \lambda \in (0, \infty) \text{ and } f(1) = 0, \ f'(1) = 1
$$

form, respectively, the Lagrangian and Eulerian subfamilies of the *Hill family of strain tensors* (cf., [3, 19]).

Table 2.1 gives some strain tensors $\mathbf{E}$ and $\mathbf{e}$ belonging to the Hill family. Note that all strain tensors given in Table 2.1 belong to the two-parameter Curnier–Rakotomanana [8] and the two-parameter power [20] strain tensor families. The strain tensors generated by the scale functions $f^{(2)}(\lambda)$, $f^{(-2)}(\lambda)$, $f^{(1)}(\lambda)$, $f^{(-1)}(\lambda)$, and $f^{(0)}(\lambda)$ also belong to the one-parameter Doyle–Ericksen [21] strain tensor family. In addition, the Hencky and Mooney strain tensors also belong to the one-parameter Itskov [2, 22] and the symmetrically physical [23] strain tensor families.

Remark 2.2 Six tensors from the Hill family of strain tensors ($\mathbf{E}^{(2)}, \mathbf{e}^{(2)}, \mathbf{E}^{(-2)}, \mathbf{e}^{(-2)}$, $\mathbf{E}^M$, and $\mathbf{e}^M$) can be defined directly in terms of the deformation gradient tensor $\mathbf{F}$ without using spectral representations (see Table 2.1). This makes them suitable for applications.

It can be shown that the Mooney strain tensors $\mathbf{E}^M$ and $\mathbf{e}^M$ approximate the Hencky strain tensors $\ln \mathbf{U}$ and $\ln \mathbf{V}$ up to third-order terms, and the tensors $\mathbf{E}^{(2)}, \mathbf{E}^{(-2)}, \mathbf{e}^{(2)}$, and $\mathbf{e}^{(-2)}$ approximate these tensors only up to second-order terms. Plots of the scale functions generating these strain tensors versus λ are given in Fig. 2.1.

Hereinafter, we assume that all tensors $\mathbf{H} \in \mathcal{T}^2$ are sufficiently smooth functions of the monotonically increasing parameter t (time), and we define the *material time derivative* (*material rate*) of the tensor $\mathbf{H}$: $\dot{\mathbf{H}} \equiv \partial \mathbf{H}/\partial t$. We introduce the *velocity vector* $\mathbf{v}$, the *velocity gradient* $\mathbf{l}$

$$
\mathbf{v} \equiv \dot{\mathbf{x}}, \quad \mathbf{l} \equiv \operatorname{grad} \mathbf{v} = \mathbf{r}_F \equiv \dot{\mathbf{F}} \cdot \mathbf{F}^{-1}, \qquad\qquad (2.7)
$$

---

[7] Further, objective Eulerian and Lagrangian tensors will be called for brevity Eulerian and Lagrangian tensors.

Table 2.1 Some strain tensors belonging to the Hill family

Generating scale function	Objectivity type	Basis-free expression	Strain tensor name
$f^{(2)}(\lambda) \equiv \frac{1}{2}(\lambda^2 - 1)$	Lagrangian	$\mathbf{E}^{(2)} \equiv \frac{1}{2}(\mathbf{U}^2 - \mathbf{I}) = \frac{1}{2}(\mathbf{F}^T \cdot \mathbf{F} - \mathbf{I})$	Green–Lagrange
	Eulerian	$\mathbf{e}^{(2)} \equiv \frac{1}{2}(\mathbf{V}^2 - \mathbf{I}) = \frac{1}{2}(\mathbf{F} \cdot \mathbf{F}^T - \mathbf{I})$	Finger
$f^{(1)}(\lambda) \equiv \lambda - 1$	Lagrangian	$\mathbf{E}^{(1)} \equiv \mathbf{U} - \mathbf{I}$	Right Biot
	Eulerian	$\mathbf{e}^{(1)} \equiv \mathbf{V} - \mathbf{I}$	Left Biot
$f^{(0)}(\lambda) \equiv \ln \lambda$	Lagrangian	$\mathbf{E}^{(0)} \equiv \ln \mathbf{U}$	Right Hencky
	Eulerian	$\mathbf{e}^{(0)} \equiv \ln \mathbf{V}$	Left Hencky
$f^{(-1)}(\lambda) \equiv 1 - \lambda^{-1}$	Lagrangian	$\mathbf{E}^{(-1)} \equiv \mathbf{I} - \mathbf{U}^{-1}$	Hill
	Eulerian	$\mathbf{e}^{(-1)} \equiv \mathbf{I} - \mathbf{V}^{-1}$	Swainger
$f^{(-2)}(\lambda) \equiv \frac{1}{2}(1 - \lambda^{-2})$	Lagrangian	$\mathbf{E}^{(-2)} \equiv \frac{1}{2}(\mathbf{I} - \mathbf{U}^{-2}) = \frac{1}{2}(\mathbf{I} - \mathbf{F}^{-1} \cdot \mathbf{F}^{-T})$	Karni–Reiner
	Eulerian	$\mathbf{e}^{(-2)} \equiv \frac{1}{2}(\mathbf{I} - \mathbf{V}^{-2}) = \frac{1}{2}(\mathbf{I} - \mathbf{F}^{-T} \cdot \mathbf{F}^{-1})$	Almansi
$f^M(\lambda) \equiv \frac{1}{4}(\lambda^2 - \lambda^{-2})$	Lagrangian	$\mathbf{E}^M \equiv \frac{1}{4}(\mathbf{U}^2 - \mathbf{U}^{-2}) = \frac{1}{4}(\mathbf{F}^T \cdot \mathbf{F} - \mathbf{F}^{-1} \cdot \mathbf{F}^{-T})$	Right Mooney
	Eulerian	$\mathbf{e}^M \equiv \frac{1}{4}(\mathbf{V}^2 - \mathbf{V}^{-2}) = \frac{1}{4}(\mathbf{F} \cdot \mathbf{F}^T - \mathbf{F}^{-T} \cdot \mathbf{F}^{-1})$	Left Mooney

Fig. 2.1 Plots of the scale functions $f(\lambda)$ generating the strain tensors given in Table 2.1: curves ##1–6 correspond to the functions $f^{(2)}(\lambda)$, $f^{(1)}(\lambda)$, $f^{(-1)}(\lambda)$, $f^{(-2)}(\lambda)$, $f^{(0)}(\lambda) = \ln \lambda$, and $f^M(\lambda)$, respectively, and the left and right vertical dashed lines correspond to $\lambda = 0.5$ and $\lambda = 2$, respectively

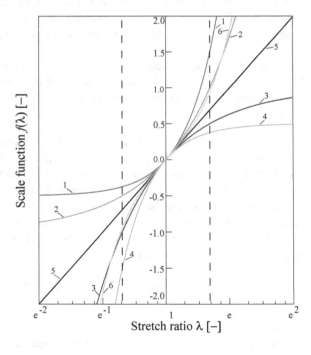

the symmetric Eulerian *stretching (strain rate) tensor* $\mathbf{d}$, the skew-symmetric *vorticity tensor* $\mathbf{w} \in \mathcal{T}^2_{skew}$,[8] and the skew-symmetric *polar spin tensor* $\boldsymbol{\omega}^R \in \mathcal{T}^2_{skew}$

$$\mathbf{l} = \mathbf{d} + \mathbf{w}, \quad \mathbf{d} \equiv \text{sym}\,\mathbf{l} = \frac{1}{2}(\mathbf{l} + \mathbf{l}^T), \quad \mathbf{w} \equiv \text{skew}\,\mathbf{l} = \frac{1}{2}(\mathbf{l} - \mathbf{l}^T), \quad \boldsymbol{\omega}^R \equiv \dot{\mathbf{R}} \cdot \mathbf{R}^T.$$
(2.8)

Introduce the Lagrangian tensor (cf., [14])

$$\mathbf{r}_U \equiv \dot{\mathbf{U}} \cdot \mathbf{U}^{-1} = \mathbf{R}^T \cdot (\mathbf{l} - \boldsymbol{\omega}^R) \cdot \mathbf{R},$$
(2.9)

and its symmetric and skew-symmetric constituents

$$\mathbf{r}_U = \mathbf{D} + \mathbf{W}, \quad \mathbf{D} \equiv \text{sym}\,\mathbf{r}_U = \frac{1}{2}(\dot{\mathbf{U}} \cdot \mathbf{U}^{-1} + \mathbf{U}^{-1} \cdot \dot{\mathbf{U}}),$$

$$\mathbf{W} \equiv \text{skew}\,\mathbf{r}_U = \frac{1}{2}(\dot{\mathbf{U}} \cdot \mathbf{U}^{-1} - \mathbf{U}^{-1} \cdot \dot{\mathbf{U}}).$$
(2.10)

We can show the validity of the relations

$$\mathbf{d} = \mathbf{R} \cdot \mathbf{D} \cdot \mathbf{R}^T, \quad \mathbf{w} = \boldsymbol{\omega}^R + \mathbf{R} \cdot \mathbf{W} \cdot \mathbf{R}^T.$$
(2.11)

The Lagrangian counterpart $\mathbf{D}$ of the tensor $\mathbf{d}$ will be called the *rotated stretching (strain rate) tensor*, and the tensor $\mathbf{W}$ will be called the *Lagrangian vorticity tensor*.

We have the following expression for the tensor $\mathbf{W}$ (see Eq. $(117)_1$ in [5])[9]:

$$\mathbf{W} = \sum_{i \neq j=1}^{m} \frac{\lambda_i - \lambda_j}{\lambda_i + \lambda_j} \mathbf{U}_i \cdot \mathbf{D} \cdot \mathbf{U}_j.$$

Substituting this expression for $\mathbf{W}$ into $(2.11)_2$, we obtain the relation between the spin tensors $\mathbf{w}$ and $\boldsymbol{\omega}^R$ (see also Eq. $(117)_2$ in [5]):

$$\mathbf{w} = \boldsymbol{\omega}^R + \sum_{i \neq j=1}^{m} \frac{\lambda_i - \lambda_j}{\lambda_i + \lambda_j} \mathbf{V}_i \cdot \mathbf{d} \cdot \mathbf{V}_j.$$
(2.12)

We can show that the following equalities hold (cf., [5, 24])[10]:

---

[8] Hereinafter, $\mathcal{T}^2_{skew} \subset \mathcal{T}^2$ denotes the set of all skew-symmetric tensors.

[9] Hereinafter, the notation $\sum_{i \neq j=1}^{m}$ denotes the summation over $i, j = 1, \ldots, m$ and $i \neq j$ and this summation is assumed to vanish when $m = 1$.

[10] Hereinafter, the notation $\overline{(\ldots)}$ means $\partial(\ldots)/\partial t$.

$$\mathbf{D} = \sum_{i=1}^{m} \overline{\ln \lambda_i} \, \mathbf{U}_i + \sum_{i \neq j=1}^{m} \mathbf{U}_i \cdot \mathbf{D} \cdot \mathbf{U}_j, \quad \mathbf{d} = \sum_{i=1}^{m} \overline{\ln \lambda_i} \, \mathbf{V}_i + \sum_{i \neq j=1}^{m} \mathbf{V}_i \cdot \mathbf{d} \cdot \mathbf{V}_j,$$

$$(2.13)$$

where the terms on the right-hand sides of these equalities are coaxial and orthogonal components to the tensors $\mathbf{U}$ and $\mathbf{V}$, respectively.

2.2 Objective Tensor Rates

It is known that the material rate of any Eulerian tensor is generally not an objective tensor. In Sect. 2.2.1 we present some (required for our study) objective corotational tensor rates of Eulerian tensors known in the literature. In Sect. 2.2.2 we consider their lesser known Lagrangian counterparts, and in Sect. 2.2.3, we establish relations between Lagrangian and Eulerian tensors and their rates.

2.2.1 Objective Rates of Eulerian Tensors

Let $\mathbf{h} \in \mathcal{T}^2$ be an Eulerian tensor. We determine objective Eulerian tensors that are the *upper and lower Oldroyd rates* of the tensor $\mathbf{h}$ [25, 26]

$$\mathbf{h}^{\sharp} = \dot{\mathbf{h}} - \mathbf{h} \cdot \mathbf{l}^T - \mathbf{l} \cdot \mathbf{h}, \quad \mathbf{h}^{\flat} = \dot{\mathbf{h}} + \mathbf{h} \cdot \mathbf{l} + \mathbf{l}^T \cdot \mathbf{h}.$$

The upper and lower Oldroyd rates can be represented in the form of the *Lie time derivatives* of the tensor $\mathbf{h}$ (see, e.g., [14, 17, 27]):

$$\mathbf{h}^{\sharp} = \mathbf{F} \cdot (\overline{\mathbf{F}^{-1} \cdot \mathbf{h} \cdot \mathbf{F}^{-T}}) \cdot \mathbf{F}^T, \quad \mathbf{h}^{\flat} = \mathbf{F}^{-T} \cdot (\overline{\mathbf{F}^T \cdot \mathbf{h} \cdot \mathbf{F}}) \cdot \mathbf{F}^{-1}. \qquad (2.14)$$

We introduce *the family of Eulerian corotational rates* of the Eulerian tensor $\mathbf{h}$ as tensors of the form

$$\mathbf{h}^{\nabla} \equiv \dot{\mathbf{h}} + \mathbf{h} \cdot \omega - \omega \cdot \mathbf{h}, \qquad (2.15)$$

associated with *the spin tensors* $\omega \in \mathcal{T}^2_{\text{skew}}$ which belong to *the family of continuous material spin tensors* [5]. This family, in turn, is a subfamily of *the family of material spin tensors* introduced by Xiao et al. [28, 29].

We introduce the tensor $\mathbf{\Psi}^E \in \mathcal{T}^{2+}_{\text{orth}}$ which is associated with the tensor ω and is determined by solving the Cauchy problem

$$\dot{\mathbf{\Psi}}^E = \omega \cdot \mathbf{\Psi}^E, \quad \mathbf{\Psi}^E = \mathbf{I} \text{ at } t = t_0. \qquad (2.16)$$

Using the tensor $\mathbf{\Psi}^E$, we can represent the tensor $\mathbf{h}^{\nabla}$ of the form (2.15) as follows:

$$\mathbf{h}^\nabla = \boldsymbol{\Psi}^E \cdot (\overline{\boldsymbol{\Psi}^{ET} \cdot \mathbf{h} \cdot \boldsymbol{\Psi}^E}) \cdot \boldsymbol{\Psi}^{ET}. \tag{2.17}$$

Some Eulerian strain tensors from the Hill family given in Table 2.1 are related to the stretching tensor $\mathbf{d}$ by the following equalities (see, e.g., [24, 30, 31])[11]:

$$\mathbf{e}^{(2)\,\sharp} = \mathbf{d}, \quad \mathbf{e}^{(-2)\,\flat} = \mathbf{d}, \quad \mathbf{e}^{(0)\,\log} = \mathbf{d}. \tag{2.18}$$

In view of expressions (2.14) and (2.17), equalities (2.18) can be written as

$$(\mathbf{e}^{(2)\,\sharp} =) \mathbf{F} \cdot (\overline{\mathbf{F}^{-1} \cdot \mathbf{e}^{(2)} \cdot \mathbf{F}^{-T}}) \cdot \mathbf{F}^T = \mathbf{d}, \tag{2.19}$$

$$(\mathbf{e}^{(-2)\,\flat} =) \mathbf{F}^{-T} \cdot (\overline{\mathbf{F}^T \cdot \mathbf{e}^{(-2)} \cdot \mathbf{F}}) \cdot \mathbf{F}^{-1} = \mathbf{d},$$

$$(\mathbf{e}^{(0)\,\log} =) \boldsymbol{\Psi}^E_{\log} \cdot (\overline{\boldsymbol{\Psi}^{ET}_{\log} \cdot \mathbf{e}^{(0)} \cdot \boldsymbol{\Psi}^E_{\log}}) \cdot \boldsymbol{\Psi}^{ET}_{\log} = \mathbf{d}.$$

2.2.2 Objective Rates of Lagrangian Tensors

Let $\mathbf{H} \in \mathcal{T}^2$ be a Lagrangian tensor. We introduce the *upper and lower Lagrangian Oldroyd's rates* of the tensor $\mathbf{H}$ (see, e.g., [14, 24]):

$$\mathbf{H}^\sharp \equiv \dot{\mathbf{H}} - \mathbf{H} \cdot \mathbf{r}_U^T - \mathbf{r}_U \cdot \mathbf{H}, \quad \mathbf{H}^\flat \equiv \dot{\mathbf{H}} + \mathbf{H} \cdot \mathbf{r}_U + \mathbf{r}_U^T \cdot \mathbf{H}. \tag{2.20}$$

Using equalities $(2.9)_1$ and (2.20), these tensor rates can be represented in the form of the Lie time derivatives of the tensor $\mathbf{H}$:

$$\mathbf{H}^\sharp = \mathbf{U} \cdot (\overline{\mathbf{U}^{-1} \cdot \mathbf{H} \cdot \mathbf{U}^{-1}}) \cdot \mathbf{U}, \quad \mathbf{H}^\flat = \mathbf{U}^{-1} \cdot (\overline{\mathbf{U} \cdot \mathbf{H} \cdot \mathbf{U}}) \cdot \mathbf{U}^{-1}. \tag{2.21}$$

We also introduce the *family of Lagrangian corotational rates* of the Lagrangian tensor $\mathbf{H}$ in the form of tensors

$$\mathbf{H}^\nabla \equiv \dot{\mathbf{H}} + \mathbf{H} \cdot \boldsymbol{\Omega} - \boldsymbol{\Omega} \cdot \mathbf{H},$$

associated with the *spin tensors* $\boldsymbol{\Omega} \in \mathcal{T}^2_{\text{skew}}$ which belong to the *family of continuous material spin tensors* [5]. For the spin tensors $\boldsymbol{\Omega}$, as for the spin tensors ω, we can introduce the tensors $\boldsymbol{\Psi}^L \in \mathcal{T}^{2+}_{\text{orth}}$ which are associated with the tensors $\boldsymbol{\Omega}$ and are determined by solving the Cauchy problem of the form (2.16):

$$\dot{\boldsymbol{\Psi}}^L = \boldsymbol{\Omega} \cdot \boldsymbol{\Psi}^L, \quad \boldsymbol{\Psi}^L = \mathbf{I} \text{ at } t = t_0. \tag{2.22}$$

---

[11] Expressions for the spin tensor $\omega^{\log}$ associated with the logarithmic corotational tensor rate $\mathbf{h}^{\log}$ of any Eulerian tensor $\mathbf{h}$ will be given in Sect. 2.3.

Using the tensor $\boldsymbol{\Psi}^L$, we obtain the following representation for $\mathbf{H}^\nabla$ of the form (2.21):

$$\mathbf{H}^\nabla = \boldsymbol{\Psi}^L \cdot (\overline{\boldsymbol{\Psi}^{LT} \cdot \mathbf{H} \cdot \boldsymbol{\Psi}^L})^{\cdot} \cdot \boldsymbol{\Psi}^{LT}. \tag{2.23}$$

Some Lagrangian strain tensors from the Hill family given in Table 2.1 are related to the Lagrangian stretching tensor $\mathbf{D}$ as follows (see, e.g., [24])[12]:

$$\mathbf{E}^{(2)\,\sharp} = \mathbf{D}, \quad \mathbf{E}^{(-2)\,\flat} = \mathbf{D}, \quad \mathbf{E}^{(0)\,\log} = \mathbf{D}. \tag{2.24}$$

In view of expressions (2.21) and (2.23), equalities (2.24) can be written as

$$(\mathbf{E}^{(2)\,\sharp} =) \, \mathbf{U} \cdot (\overline{\mathbf{U}^{-1} \cdot \mathbf{E}^{(2)} \cdot \mathbf{U}^{-1}})^{\cdot} \cdot \mathbf{U} = \mathbf{D},$$

$$(\mathbf{E}^{(-2)\,\flat} =) \, \mathbf{U}^{-1} \cdot (\overline{\mathbf{U} \cdot \mathbf{E}^{(-2)} \cdot \mathbf{U}})^{\cdot} \cdot \mathbf{U}^{-1} = \mathbf{D},$$

$$(\mathbf{E}^{(0)\,\log} =) \, \boldsymbol{\Psi}^L_{\log} \cdot (\overline{\boldsymbol{\Psi}^{LT}_{\log} \cdot \mathbf{E}^{(0)} \cdot \boldsymbol{\Psi}^L_{\log}})^{\cdot} \cdot \boldsymbol{\Psi}^{LT}_{\log} = \mathbf{D}.$$

2.2.3   Relationship Between Lagrangian and Eulerian Tensors and Their Rates

Let $\mathbf{h}$, $\mathbf{H} \in \mathcal{T}^2$ be Eulerian and Lagrangian tensors, respectively. We will call these tensors *objective counterparts* of each other if they are related by the equalities

$$\mathbf{h} = \mathbf{R} \cdot \mathbf{H} \cdot \mathbf{R}^T \quad \Leftrightarrow \quad \mathbf{H} = \mathbf{R}^T \cdot \mathbf{h} \cdot \mathbf{R}. \tag{2.25}$$

If $\mathbf{h}$, $\mathbf{H} \in \mathcal{T}^2$ are Eulerian and Lagrangian counterparts of each other, it can be shown that the convective tensor rates $\mathbf{h}^\sharp$, $\mathbf{h}^\flat$ and $\mathbf{H}^\sharp$, $\mathbf{H}^\flat$ are also objective counterparts of each other.

Proposition 2.1 *Let $\mathbf{h}$, $\mathbf{H} \in \mathcal{T}^2$ be Eulerian and Lagrangian tensors, respectively, which are objective counterparts of each other. Then the necessary and sufficient condition for the corotational rates $\mathbf{h}^\nabla$ and $\mathbf{H}^\nabla$ of the tensors $\mathbf{h}$ and $\mathbf{H}$ to be objective counterparts of each other is given by*

$$\boldsymbol{\omega} = \boldsymbol{\omega}^R + \mathbf{R} \cdot \boldsymbol{\Omega} \cdot \mathbf{R}^T, \tag{2.26}$$

where $\boldsymbol{\omega}$, $\boldsymbol{\Omega} \in \mathcal{T}^2_{\text{skew}}$ are spin tensors associated with the corotational tensor rates $\mathbf{h}^\nabla$ and $\mathbf{H}^\nabla$, respectively.

Proof (i) Let $\mathbf{h}^\nabla$ and $\mathbf{H}^\nabla$ be objective counterparts of each other. Then (2.25) leads to the equality

---

[12] Expressions for the spin tensor $\boldsymbol{\Omega}^{\log}$ associated with the logarithmic corotational tensor rate $\mathbf{H}^{\log}$ of any Lagrangian tensor $\mathbf{H}$ will be given in Sect. 2.3.

$$\mathbf{h}^{\nabla} = \mathbf{R} \cdot \mathbf{H}^{\nabla} \cdot \mathbf{R}^T. \tag{2.27}$$

Using the definitions of the corotational rates $\mathbf{h}^{\nabla}$ and $\mathbf{H}^{\nabla}$ above, from (2.27) we obtain

$$\dot{\mathbf{h}} + \mathbf{h} \cdot \boldsymbol{\omega} - \boldsymbol{\omega} \cdot \mathbf{h} = \mathbf{R} \cdot \dot{\mathbf{H}} \cdot \mathbf{R}^T + \mathbf{R} \cdot \mathbf{H} \cdot \mathbf{R}^T \cdot \mathbf{R} \cdot \boldsymbol{\Omega} \cdot \mathbf{R}^T - \mathbf{R} \cdot \boldsymbol{\Omega} \cdot \mathbf{R}^T \cdot \mathbf{R} \cdot \mathbf{H} \cdot \mathbf{R}^T. \tag{2.28}$$

Using $(2.8)_4$ and (2.25), we transform (2.28) to the equality

$$\dot{\mathbf{h}} + \mathbf{h} \cdot \boldsymbol{\omega} - \boldsymbol{\omega} \cdot \mathbf{h} = \dot{\mathbf{h}} + \mathbf{h} \cdot (\boldsymbol{\omega}^R + \mathbf{R} \cdot \boldsymbol{\Omega} \cdot \mathbf{R}^T) - (\boldsymbol{\omega}^R + \mathbf{R} \cdot \boldsymbol{\Omega} \cdot \mathbf{R}^T) \cdot \mathbf{h}. \tag{2.29}$$

Since $\mathbf{h}$ and $\boldsymbol{\omega}$ are arbitrary, equality (2.29) leads to (2.26).

(ii) Let equality (2.26) hold. Applying the manipulations in reverse order, from (2.29) we obtain (2.27). □

Proposition 2.2 *Let*

(i) $\boldsymbol{\omega} \in \mathcal{T}_{\text{skew}}^2$ *be some skew-symmetric tensor and let* $\boldsymbol{\Psi} \in \mathcal{T}_{\text{orth}}^{2+}$ *be the rotation tensor which is associated with the spin tensor* $\boldsymbol{\omega}$*, and is determined by solving the Cauchy problem*

$$\dot{\boldsymbol{\Psi}} = \boldsymbol{\omega} \cdot \boldsymbol{\Psi}, \quad \boldsymbol{\Psi} = \mathbf{I} \text{ at } t = t_0; \tag{2.30}$$

(ii) the tensor $\boldsymbol{\omega}$ *can be represented as the sum of two tensors* $\boldsymbol{\omega}_1, \boldsymbol{\omega}_2 \in \mathcal{T}_{\text{skew}}^2$

$$\boldsymbol{\omega} = \boldsymbol{\omega}_1 + \boldsymbol{\omega}_2. \tag{2.31}$$

Then the tensor $\boldsymbol{\Psi}$ *can be represented as the inner product of two tensors* $\boldsymbol{\Psi}_1, \boldsymbol{\Psi}_2 \in \mathcal{T}_{\text{orth}}^{2+}$

$$\boldsymbol{\Psi} = \boldsymbol{\Psi}_1 \cdot \boldsymbol{\Psi}_2, \tag{2.32}$$

where the tensor $\boldsymbol{\Psi}_1$ *is associated with the spin tensor* $\boldsymbol{\omega}_1$ *and is determined by solving the Cauchy problem*

$$\dot{\boldsymbol{\Psi}}_1 = \boldsymbol{\omega}_1 \cdot \boldsymbol{\Psi}_1, \quad \boldsymbol{\Psi}_1 = \mathbf{I} \text{ at } t = t_0; \tag{2.33}$$

and the tensor $\boldsymbol{\Psi}_2$ *is associated with the spin tensor* $\tilde{\boldsymbol{\omega}}_2 \in \mathcal{T}_{\text{skew}}^2$

$$\tilde{\boldsymbol{\omega}}_2 \equiv \boldsymbol{\Psi}_1^T \cdot \boldsymbol{\omega}_2 \cdot \boldsymbol{\Psi}_1, \tag{2.34}$$

and is determined by solving the Cauchy problem

$$\dot{\boldsymbol{\Psi}}_2 = \tilde{\boldsymbol{\omega}}_2 \cdot \boldsymbol{\Psi}_2, \quad \boldsymbol{\Psi}_2 = \mathbf{I} \text{ at } t = t_0. \tag{2.35}$$

Proof Let the tensor $\boldsymbol{\Psi}$ have the form (2.32). We show that in this case, equation $(2.30)_1$ is identically satisfied. Using expression (2.32) and equalities $(2.33)_1$ and $(2.35)_1$, we transform the l.h.s. of $(2.30)_1$:

The l.h.s. of Eq. $(2.30)_1 = \dot{\Psi}_1 \cdot \Psi_2 + \Psi_1 \cdot \dot{\Psi}_2 = \omega_1 \cdot \Psi_1 \cdot \Psi_2 + \Psi_1 \cdot \tilde{\omega}_2 \cdot \Psi_2.$

$$(2.36)$$

Using (2.34), from (2.31), (2.32), and (2.36), we obtain

$$\text{The l.h.s. of Eq. } (2.30)_1 = \omega_1 \cdot \Psi + \omega_2 \cdot \Psi = \omega \cdot \Psi,$$

whence it follows that the l.h.s. of $(2.30)_1$ is equal to the r.h.s. of $(2.30)_1$; i.e., equation $(2.30)_1$ is identically satisfied. It is easy to check that for the tensor Ψ represented in the form (2.32), the initial conditions $(2.30)_2$ are satisfied if the initial conditions $(2.33)_2$ and $(2.35)_2$ are satisfied. $\qquad\square$

Proposition 2.3 *Suppose that*

(i) $\mathbf{h}$, $\mathbf{H} \in \mathcal{T}^2$ *are Eulerian and Lagrangian tensors, respectively, that are objective counterparts of each other;*

(ii) ω, $\Omega \in \mathcal{T}^2_{\text{skew}}$ *are spin tensors associated with corotational rates $\mathbf{h}^\nabla$ and $\mathbf{H}^\nabla$, respectively, that are also objective counterparts of each other;*

(iii) Ψ^E, Ψ^L, *and $\mathbf{R}$ are rotation tensors associated with the spin tensors ω, Ω, and ω^R.*

Then the tensors Ψ^E, Ψ^L, and $\mathbf{R}$ are related by the equality

$$\Psi^E = \mathbf{R} \cdot \Psi^L. \tag{2.37}$$

Proof of this proposition follows directly from Propositions 2.1 and 2.2. $\qquad\square$

Exercise 1.36 in [2] leads to the following proposition.

Proposition 2.4 *Let $\mathbf{A}$ be an arbitrary second-order tensor. Then the following equality is valid:*

$$\exp(\Psi \cdot \mathbf{A} \cdot \Psi^T) = \Psi \cdot \exp(\mathbf{A}) \cdot \Psi^T \quad \forall \Psi \in \mathcal{T}^{2+}_{\text{orth}}.$$

2.3   Hooke-Like Isotropic Hypoelasticity Models Based on Corotational Stress Rates

Choosing the configuration of the body $\mathfrak{B}$ at the initial time t_0 as the reference configuration (see, e.g., [32]), we consider CRs for isotropic generalized Hooke-like hypoelastic material models in the following Lagrangian and Eulerian forms:

$$\bar{\tau}^\Omega = \lambda \operatorname{tr}\mathbf{D}\,\mathbf{I} + 2\mu\mathbf{D}, \quad \tau^\omega = \lambda \operatorname{tr}\mathbf{d}\,\mathbf{I} + 2\mu\mathbf{d}.$$

Hereinafter, λ and μ are the *Lamé parameters*, $\tau \equiv J\sigma$ (σ is the Cauchy stress tensor) and $\bar{\tau} \equiv \mathbf{R}^T \cdot \tau \cdot \mathbf{R}$ are the *standard/rotated Kirchhoff stress tensors*, respectively.

The tensors $\bar{\tau}^{\Omega}$ and τ^{ω} are the Lagrangian and Eulerian corotational rates of the tensors $\bar{\tau}$ and τ (objective counterparts of each other), respectively; they are associated with the spin tensors Ω and ω, are objective counterparts of each other, and have the form

$$\bar{\tau}^{\Omega} \equiv \dot{\bar{\tau}} + \bar{\tau} \cdot \Omega - \Omega \cdot \bar{\tau}, \quad \tau^{\omega} \equiv \dot{\tau} + \tau \cdot \omega - \omega \cdot \tau.$$

Since, like the tensors $\bar{\tau}^{\Omega}$ and τ^{ω}, the tensors τ and $\bar{\tau}$ are objective counterparts of each other, it follows from Propositions 2.1 and 2.3 that the spin tensors ω and Ω are related by equality (2.26) and their associated rotation tensors Ψ^{E} and Ψ^{L} are related by equality (2.37). Let the tensors Ψ^{L} and Ψ^{E} associated with the spin tensors Ω and ω be determined by solving problems (2.22) and (2.16). Then, in view of (2.23) and (2.17), the tensors $\bar{\tau}^{\Omega}$ and τ^{ω} can be represented as

$$\bar{\tau}^{\Omega} = \Psi^{L} \cdot (\overline{\Psi^{LT} \cdot \bar{\tau} \cdot \Psi^{L}})\dot{} \cdot \Psi^{LT}, \quad \tau^{\omega} = \Psi^{E} \cdot (\overline{\Psi^{ET} \cdot \tau \cdot \Psi^{E}})\dot{} \cdot \Psi^{ET}. \quad (2.38)$$

For a given value of the Cauchy initial stress σ^{0} and a given law of motion $\mathbf{x} = \mathbf{x}(\mathbf{X}, t)$, the determination of the tensors $\bar{\tau}$ and τ requires the solution of the Cauchy problems

$$\bar{\tau}^{\Omega} = \lambda \operatorname{tr}\mathbf{D}\,\mathbf{I} + 2\mu\mathbf{D}, \quad \bar{\tau} = \sigma^{0} \text{ at time } t_0, \quad (2.39)$$

and

$$\tau^{\omega} = \lambda \operatorname{tr}\mathbf{d}\,\mathbf{I} + 2\mu\mathbf{d}, \quad \tau = \sigma^{0} \text{ at time } t_0, \quad (2.40)$$

respectively. Using expressions (2.38) and the equality (see, e.g., [24])

$$\operatorname{tr}\mathbf{D} = \operatorname{tr}\mathbf{d} = \overline{\ln J},$$

we can represent the solutions of problems (2.39) and (2.40) in the form [33]

$$\bar{\tau} = \lambda \ln J\,\mathbf{I} + \bar{\tau}_n, \quad \tau = \lambda \ln J\,\mathbf{I} + \tau_n, \quad (2.41)$$

where the tensors $\bar{\tau}_n$ and τ_n are the solutions of the Cauchy problems (omitting the subscript n for brevity)

$$\bar{\tau}^{\Omega} = 2\mu\mathbf{D}, \quad \bar{\tau} = \sigma^{0} \text{ at time } t_0, \quad (2.42)$$

and

$$\tau^{\omega} = 2\mu\mathbf{d}, \quad \tau = \sigma^{0} \text{ at time } t_0. \quad (2.43)$$

Note that the determination of the tensors $\bar{\tau}_n$ and τ_n generally requires the numerical solutions of problems (2.42) and (2.43).

Next we explore the spin tensors Ω and ω from the family of continuous material spin tensors in two alternative versions (see, e.g., [5, 30]):

(i) *r-version*

$$\mathbf{\Omega} = \mathbf{\Psi}_r(\mathbf{U}, \mathbf{D}), \quad \omega = \omega^R + \mathbf{\Psi}_r(\mathbf{V}, \mathbf{d}). \tag{2.44}$$

Here $\mathbf{\Psi}_r(\mathbf{U}, \mathbf{D}) \in \mathcal{T}_{\text{skew}}^2$ is an isotropic tensor function of the arguments $\mathbf{U}$ and $\mathbf{D}$ which is linear in $\mathbf{D}$,[13]

$$\mathbf{\Psi}_r(\mathbf{U}, \mathbf{D}) \equiv \sum_{i \neq j=1}^{m} r_{ij} \mathbf{U}_i \cdot \mathbf{D} \cdot \mathbf{U}_j, \tag{2.45}$$

where we used the notations

$$r_{ij} \equiv r(\lambda_i, \lambda_j), \quad r_{ji} = -r_{ij}.$$

The properties of the function $r(x, y)$ are discussed in detail in [28, 29] (see also [5]).

(ii) *g-version*

$$\mathbf{\Omega} = \mathbf{W} + \mathbf{\Psi}_g(\mathbf{U}, \mathbf{D}), \quad \omega = \mathbf{w} + \mathbf{\Psi}_g(\mathbf{V}, \mathbf{d}). \tag{2.46}$$

Here $\mathbf{\Psi}_g(\mathbf{U}, \mathbf{D}) \in \mathcal{T}_{\text{skew}}^2$ is an isotropic tensor function of the tensors $\mathbf{U}$ and $\mathbf{D}$ which is linear in $\mathbf{D}$,

$$\mathbf{\Psi}_g(\mathbf{U}, \mathbf{D}) \equiv \sum_{i \neq j=1}^{m} g_{ij} \mathbf{U}_i \cdot \mathbf{D} \cdot \mathbf{U}_j,$$

where we used the notations

$$g_{ij} \equiv g(\lambda_i, \lambda_j) \equiv r(\lambda_i, \lambda_j) - \frac{\lambda_i - \lambda_j}{\lambda_i + \lambda_j}, \quad g_{ji} = -g_{ij}. \tag{2.47}$$

Below we give expressions for the quantities g_{ij} and r_{ij} generating some spin tensors from the family of continuous material spin tensors and the classical corotational rates associated with them $(i, j = 1, \ldots, m)$:

- *The Zaremba–Jaumann stress rates* $\bar{\boldsymbol{\tau}}^{ZJ}$ and $\boldsymbol{\tau}^{ZJ}$ associated with the spin tensors $\mathbf{W}$ and $\mathbf{w}$

$$g_{ij}^{ZJ} = 0; \quad r_{ij}^{ZJ} \equiv \frac{\lambda_i - \lambda_j}{\lambda_i + \lambda_j}.$$

- The material rate $\dot{\bar{\boldsymbol{\tau}}}$ and *the Green–Naghdi stress rate* $\boldsymbol{\tau}^{GN}$ associated with the spin tensors $(\mathbf{\Omega} =) \mathbf{0}$ and ω^R

---

[13] The isotropic tensor functions $\mathbf{\Psi}_r(\mathbf{V}, \mathbf{d})$ and $\mathbf{\Psi}_g(\mathbf{V}, \mathbf{d})$ are derived from the isotropic tensor functions $\mathbf{\Psi}_r(\mathbf{U}, \mathbf{D})$ and $\mathbf{\Psi}_g(\mathbf{U}, \mathbf{D})$ by replacing the Lagrangian tensor arguments $\mathbf{U}$ and $\mathbf{D}$ with their Eulerian counterparts $\mathbf{V}$ and $\mathbf{d}$, respectively.

$$g_{ij}^{GN} = \frac{\lambda_j - \lambda_i}{\lambda_i + \lambda_j}; \quad r_{ij}^{GN} = 0.$$

- The *logarithmic stress rates* $\bar{\tau}^{\log}$ and $\tau^{\log}$ associated with the spin tensor $\boldsymbol{\Omega}^{\log}$ and $\omega^{\log}$

$$g_{ij}^{\log} \equiv \frac{\lambda_i^2 + \lambda_j^2}{\lambda_j^2 - \lambda_i^2} + \frac{1}{\ln \lambda_i - \ln \lambda_j}; \quad r_{ij}^{\log} \equiv \frac{2\lambda_i \lambda_j}{\lambda_j^2 - \lambda_i^2} + \frac{1}{\ln \lambda_i - \ln \lambda_j}.$$

We consider the following CRs for a Hooke-like hypoelastic material model that does not belong to the family of models with CRs considered above

$$\boldsymbol{\sigma}^H = \lambda \operatorname{tr}\mathbf{d}\,\mathbf{I} + 2\mu\mathbf{d}, \quad \boldsymbol{\sigma} = \boldsymbol{\sigma}^0 \text{ at time } t_0, \tag{2.48}$$

where $\boldsymbol{\sigma}^H$ is the *Hill stress rate* (cf., [34])[14]

$$\boldsymbol{\sigma}^H \equiv \dot{\boldsymbol{\sigma}} + \boldsymbol{\sigma} \cdot \mathbf{w} - \mathbf{w} \cdot \boldsymbol{\sigma} + \boldsymbol{\sigma}\operatorname{tr}\mathbf{d}\,(= \boldsymbol{\sigma}^{ZJ} + \boldsymbol{\sigma}\operatorname{tr}\mathbf{d}).$$

Remark 2.3 In traditional CRs for classical hypoelasticity (see, e.g., [4]), the l.h.s. of $(2.48)_1$ contains the Zaremba–Jaumann stress rate $\boldsymbol{\sigma}^{ZJ}$ instead of the Hill stress rate $\boldsymbol{\sigma}^H$. However, the use of this material model generally leads to a loss of symmetry in rate formulations of the solid mechanics equations (cf., [37]).

Remark 2.4 Some authors use the hypoelasticity model with CRs (2.40) based on the stress rate $\boldsymbol{\tau}^{ZJ}$ on the l.h.s. of this equation (see, e.g., [38]), and other authors use the hypoelasticity model with CRs $(2.48)_1$ (see, e.g., [39]) to take into account the elastic constituent in CRs for additive elasto-plasticity. Therefore, in this paper, we consider both of the hypoelasticity models. Note that for isochoric deformation (e.g., simple shear), these two hypoelasticity models coalesce, but in simulations of deformation with a marked change in volume, these models can lead to significantly different values of the Cauchy stress tensor components (see, e.g., Fig. 6.2, which shows plots of Cauchy stresses versus engineering strain obtained by solving the simple extension problem using both material models).

Although CRs $(2.48)_1$ do not require the introduction of any preferred reference configuration of the body, for the integration of CRs $(2.48)_1$ with given initial stresses $\boldsymbol{\sigma}^0$ (at time t_0) using the reference configuration at time t_0, the following equality holds (see, e.g., [35]):

$$\boldsymbol{\tau}^{ZJ} = J\boldsymbol{\sigma}^H. \tag{2.49}$$

---

[14] Sometimes, the Hill stress rate is called the *Biezeno–Hencky stress rate* (cf., [35]). Following [36], we note that an expression for this stress rate was obtained by Hencky as early as in 1929—in the period when he collaborated with Biezeno—however, an unfortunate error in the sign in determining the vorticity tensor makes his expression for the stress rate non-objective. In 1933 and 1947, Fromm corrected this Hencky's error in solving a specific problem (cf., [36]), and in 1958, an objective expression for this stress rate was proposed by Hill.

Using (2.49), for integrating CRs of the form (2.48)$_1$, we obtain the Lagrangian version of the Cauchy problem

$$\bar{\tau}^{ZJ} = \lambda J \operatorname{tr}\mathbf{D}\,\mathbf{I} + 2\mu J\mathbf{D}, \quad \bar{\tau} = \sigma^0 \text{ at time } t_0, \tag{2.50}$$

and the Eulerian version of this problem

$$\tau^{ZJ} = \lambda J \operatorname{tr}\mathbf{d}\,\mathbf{I} + 2\mu J\mathbf{d}, \quad \tau = \sigma^0 \text{ at time } t_0. \tag{2.51}$$

By analogy with the representations of solutions of problems (2.39) and (2.40) in the form of sums (2.41), we represent the solutions of problems (2.50) and (2.51) in the form

$$\bar{\tau} = \lambda(J-1)\mathbf{I} + \bar{\tau}_n, \quad \tau = \lambda(J-1)\mathbf{I} + \tau_n,$$

where the tensors $\bar{\tau}_n$ and τ_n are the solutions of the Cauchy problems (omitting the subscript n for brevity)

$$\bar{\tau}^{ZJ} = 2\mu J\mathbf{D}, \quad \bar{\tau} = \sigma^0 \text{ at time } t_0, \tag{2.52}$$

and

$$\tau^{ZJ} = 2\mu J\mathbf{d}, \quad \tau = \sigma^0 \text{ at time } t_0, \tag{2.53}$$

respectively.

Our next objective is to develop enhanced objective algorithms for the numerical solution of problems (2.42), (2.43), (2.52), and (2.53). We first make a classification of the considered material models based on the properties of kinematic tensors used in CRs: material models generated by kinematic tensors independent of the choice of body reference configuration (cf., [4, 40, 41]) will be called *classical material models* (*C-models*), and material models generated by kinematic tensors dependent on this choice (cf., [42]) will be called *generalized material models* (*G-models*). Since the kinematic tensors $\mathbf{w}$ and $\mathbf{d}$ do not depend on the choice of reference configuration, the material models generated by the Zaremba–Jaumann and Hill stress rates are included in C-models,[15] and the remaining models (including material models based on the Green–Naghdi and logarithmic stress rates) are included in G-models. In Chap. 6, using the proposed classification of hypoelasticity models, we show that when choosing among algorithms exactly reproducing SRBMs, it is preferable to use

---

[15] For CRs of classical hypoelasticity (cf., [4, 40, 41]) dependence on the selected reference config-uration is not allowed. Strictly speaking, of the models considered above, only the material model based on the Hill stress rate σ^H belongs to material models with such CRs, and CRs for the material model based on the Zaremba–Jaumann stress rate τ^{ZJ} depend on the choice of reference configu-ration by means of the scalar quantity J ($\tau = J\sigma$) (note that this weak dependence appears only for nonisochoric deformation). For the algorithms considered in this paper, independence from the cho-sen reference configuration is more important for the kinematic tensors $\mathbf{w}$ and $\mathbf{d}$ than for the stress tensor τ. Therefore, in this book, we include the material model based on the Zaremba–Jaumann stress rate τ^{ZJ} in the family of C-models.

strong incrementally objective algorithms for C-models (see Chap. 4) and absolutely objective algorithms for G-models (see Chap. 5).

Comparing the formulations of problems (2.42) and (2.43) with the formulations of problems (2.52) and (2.53), we note that under the assumption $J = 1$, the second two problems reduce to the first two. Therefore, further we develop objective algorithms for the numerical solution of problems (2.42) and (2.43), noting that algorithms for the numerical solution of problems (2.52) and (2.53) are only slight modifications of former ones.

References

1. A. Bertram, *Elasticity and Plasticity of Large Deformations*, 4th edn. (Springer, Cham, 2021)
2. M. Itskov, *Tensor Algebra and Tensor Analysis for Engineers (with Applications to Continuum Mechanics)*, 5th edn. (Springer, Switzerland, 2019)
3. R.W. Ogden, *Non-linear Elastic Deformations* (Ellis Horwood, Chichester, 1984)
4. C. Truesdell, W. Noll, in *Handbuch der Physik, Vol. III/3*, ed. by S. Flügge (Springer, Berlin, 1965), pp. 1–602
5. S.N. Korobeynikov, Acta Mech. **216**, 301 (2011). https://doi.org/10.1007/s00707-010-0369-7
6. K. Hashiguchi, Y. Yamakawa, *Introduction to Finite Strain Theory for Continuum Elasto-Plasticity* (Wiley, Hoboken, 2013)
7. C.P. Luehr, M.B. Rubin, Comput. Methods Appl. Mech. Eng. **84**, 243 (1990). https://doi.org/10.1016/0045-7825(90)90078-Z
8. A. Curnier, L. Rakotomanana, Eng. Trans. **39**, 461 (1991)
9. S.N. Korobeynikov, Acta Mech. **229**, 1061 (2018). https://doi.org/10.1007/s00707-017-1972-7
10. K.J. Bathe, *Finite Element Procedures* (Prentice Hall, Upper Saddle River, New Jersey, 1996)
11. S. Roy, A.F. Fossum, R.J. Dexter, Int. J. Eng. Sci. **30**, 119 (1992). https://doi.org/10.1016/0020-7225(92)90045-I
12. A. Hoger, D.E. Carlson, Quart. Appl. Math. **42**, 113 (1984). https://doi.org/10.1090/qam/736511
13. N.H. Scott, J. Elast. **141**, 363 (2020). https://doi.org/10.1007/s10659-020-09780-x
14. S.N. Korobeynikov, J. Elast. **93**, 105 (2008). https://doi.org/10.1007/s10659-008-9166-0
15. I.S. Liu, J. Elast. **71**, 73 (2003). https://doi.org/10.1023/B:ELAS.0000005548.36767.e7
16. A.I. Murdoch, J. Elast. **60**, 233 (2000). https://doi.org/10.1023/A:1011049615372
17. G.A. Holzapfel, *Nonlinear Solid Mechanics: A Continuum Approach for Engineering* (Wiley, Chichester, 2000)
18. P. Haupt, C. Tsakmakis, Contin. Mech. Thermodyn. **1**, 165 (1989). https://doi.org/10.1007/BF01171378
19. R. Hill, in *Advances in Applied Mechanics*, ed. by C.-S. Yih (Academic, New York, 1979), pp.1–75
20. H. Darijani, R. Naghdabadi, Int. J. Eng. Sci. **48**, 223 (2010). https://doi.org/10.1016/j.ijengsci.2009.08.006
21. T.C. Doyle, J.L. Ericksen, in *Advances in Applied Mechanics*, vol. 4, ed. by H. Dryden, T. von Karman (Academic, New York, 1956), pp.53–115
22. M. Itskov, Mech. Res. Commun. **31**, 507 (2004). https://doi.org/10.1016/j.mechrescom.2004.02.006
23. S.N. Korobeynikov, J. Elast. **136**, 159 (2019). https://doi.org/10.1007/s10659-018-9699-9
24. S.N. Korobeynikov, J. Elast. **143**, 147 (2021). https://doi.org/10.1007/s10659-020-09808-2
25. J.G. Oldroyd, Proc. R. Soc. A **200**, 523 (1950). https://doi.org/10.1098/rspa.1950.0035

26. J.G. Oldroyd, Proc. R. Soc. A **245**, 278 (1958). https://doi.org/10.1098/rspa.1958.0083
27. J.E. Marsden, T.J.R. Hughes, *Mathematical Foundations of Elasticity* (Prentice-Hall, Englewood Cliffs, 1983)
28. H. Xiao, O.T. Bruhns, A. Meyers, Int. J. Solids Struct. **35**, 4001 (1998). https://doi.org/10.1016/S0020-7683(97)00267-9
29. H. Xiao, O.T. Bruhns, A. Meyers, J. Elast. **52**, 1 (1998). https://doi.org/10.1023/A:1007570827614
30. S.N. Korobeynikov, Arch. Appl. Mech. **90**, 313 (2020). https://doi.org/10.1007/s00419-019-01611-3
31. O.T. Bruhns, A. Meyers, H. Xiao, Proc. R. Soc. A **460**, 909 (2004). https://doi.org/10.1098/rspa.2003.1184
32. M. Kleiber, P. Kowalczyk, *Introduction to Nonlinear Thermomechanics of Solids* (Springer, Switzerland, 2016)
33. A.M. Lush, G. Weber, L. Anand, Int. J. Plast. **5**, 521 (1989). https://doi.org/10.1016/0749-6419(89)90012-0
34. R. Hill, J. Mech. Phys. Solids **6**, 236 (1958). https://doi.org/10.1016/0022-5096(58)90029-2
35. W. Ji, A.M. Waas, Z.P. Bažant, J. Appl. Mech. **80**, 041024 (2013). https://doi.org/10.1115/1.4007828
36. R.I. Tanner, E. Tanner, Rheol. Acta **42**, 93 (2003). https://doi.org/10.1007/s00397-002-0259-6
37. R. Hill, J. Mech. Phys. Solids **7**, 209 (1959). https://doi.org/10.1016/0022-5096(59)90007-92
38. E.A. de Souza Neto, D. Peric, D.J.R. Owen, *Computational Methods for Plasticity: Theory and Applications* (Wiley, Chichester, 2008)
39. R. McMeeking, J. Rice, Int. J. Solids Struct. **11**, 601 (1975). https://doi.org/10.1016/0020-7683(75)90033-5
40. C. Truesdell, Commun. Pure Appl. Math. **8**, 123 (1955). https://doi.org/10.1002/cpa.3160080109
41. C. Truesdell, J. Ration. Mech. Anal. **4**, 83 (1955)
42. A.E. Green, B.C. McInnis, P. R. Soc. Edinb. A **67**, 220 (1967). https://doi.org/10.1017/S0080454100008074

Chapter 3
Incremental Tensors and the Incremental Objectivity of Tensors

Abstract When integrating hypoelastic CRs numerically, instead of the original rate formulations of the equations in question, one should use their incremental counterparts. To maintain the objectivity property of the original time-continuous tensors and their rates for their incremental counterparts, in Sect. 3.1 we introduce Definition 3.1 of the incremental objectivity of Eulerian tensors. In Sect. 3.2 we introduce some incrementally objective kinematic tensors, which are then used in Chaps. 4 and 5 to develop objective algorithms for integrating the CRs in question.

3.1 Definition of the Incremental Objectivity of Tensors

The time interval $[t_0, T]$ of integration of hypoelastic CRs can be represented as the union of N incremental intervals $(t_N \equiv T)$

$$[t_0, T] = [t_0, t_1] \cup [t_1, t_2] \cup \ldots \cup [t_{N-1}, t_N].$$

As is common in computational mechanics, all tensors are defined only at discrete times $t_0, t_1, \ldots, t_N$. For example, the position vector $\mathbf{x}$ and the vector $\mathbf{c}$ in (2.5) are defined at discrete times as sequences of discrete quantities

$$^{t_0}\mathbf{x}(\equiv \mathbf{X}), \, ^{t_1}\mathbf{x}, \ldots, \, ^{t_{N-1}}\mathbf{x}, \, ^{t_N}\mathbf{x},$$
$$^{t_0}\mathbf{c}(\equiv \mathbf{c}_0), \, ^{t_1}\mathbf{c}, \ldots, \, ^{t_{N-1}}\mathbf{c}, \, ^{t_N}\mathbf{c},$$

and the tensor $\mathbf{Q}$ is defined at discrete times as a sequence of proper orthogonal tensors of the form

$$^{t_0}\mathbf{Q}(\equiv \mathbf{Q}_0), \, ^{t_1}\mathbf{Q}, \ldots, \, ^{t_{N-1}}\mathbf{Q}, \, ^{t_N}\mathbf{Q}.$$

Let time t correspond to any time t_n $(n = 0, \ldots, N - 1)$; we define the *time step* $\Delta t \equiv t_{n+1} - t_n$, and the *incremental Euclidean transformations* (IETs)

$$^{t+\Delta t}\mathbf{x}^* \equiv \, ^{t+\Delta t}\mathbf{Q} \cdot \, ^{t+\Delta t}\mathbf{x} + \, ^{t+\Delta t}\mathbf{c}, \quad ^{t}\mathbf{x}^* \equiv \, ^{t}\mathbf{Q} \cdot \, ^{t}\mathbf{x} + \, ^{t}\mathbf{c}. \tag{3.1}$$

© The Author(s), under exclusive license to Springer Nature Switzerland AG 2023   25
S. Korobeynikov and A. Larichkin, *Objective Algorithms for Integrating Hypoelastic Constitutive Relations Based on Corotational Stress Rates*, SpringerBriefs in Continuum Mechanics, https://doi.org/10.1007/978-3-031-29632-1_3

Definition 3.1 A tensor $\mathbf{h} \in \mathcal{T}^2$ is called an *incrementally objective tensor* if it changes under IETs (3.1) according to the laws

$$\mathbf{h}^* = {}^{t+\Delta t}\mathbf{Q} \cdot \mathbf{h} \cdot {}^{t+\Delta t}\mathbf{Q}^T \quad \textit{incrementally Eulerian,} \tag{3.2}$$

$$\mathbf{h}^* = {}^{t}\mathbf{Q} \cdot \mathbf{h} \cdot {}^{t}\mathbf{Q}^T \qquad \textit{incrementally Lagrangian,}$$

$$\mathbf{h}^* = {}^{t+\Delta t}\mathbf{Q} \cdot \mathbf{h} \cdot {}^{t}\mathbf{Q}^T \quad \textit{incrementally Eulerian–Lagrangian,}$$

$$\mathbf{h}^* = {}^{t}\mathbf{Q} \cdot \mathbf{h} \cdot {}^{t+\Delta t}\mathbf{Q}^T \quad \textit{incrementally Lagrangian–Eulerian.}$$

Remark 3.1 To distinguish between the types of objectivities introduced in Definitions 2.1 and 3.1 in cases where confusion is possible, we call tensors that are objective in accordance with Definition 2.1 *absolutely objective tensors.*

It follows from $(2.6)_1$ and $(3.2)_{1,2}$ that the Eulerian tensor $\mathbf{h}$ is incrementally Eulerian in the time interval $[t, t + \Delta t]$ if it is defined at time $t + \Delta t$ (denote this tensor as ${}^{t+\Delta t}\mathbf{h}$) and it is incrementally Lagrangian for the same increment if it is defined at time t (denote this tensor as ${}^{t}\mathbf{h}$). Lagrangian tensors are generally not incrementally objective tensors (except in the time interval $[t_0, t_0 + \Delta t]$).

3.2   Incrementally Objective Kinematic Tensors

Consider any Eulerian–Lagrangian tensor $\mathbf{L}$ (e.g., the tensor $\mathbf{L} = \mathbf{F}$). It can be shown that the tensor

$$^{\Delta t}\mathbf{L} \equiv {}^{t+\Delta t}\mathbf{L} \cdot {}^{t}\mathbf{L}^{-1} \tag{3.3}$$

is incremental Eulerian–Lagrangian. Accordingly, the tensor $^{\Delta t}\mathbf{L}^T$ is incremental Lagrangian–Eulerian.

According to (3.3), we define the incrementally Eulerian–Lagrangian tensors

$$^{\Delta t}\mathbf{F} \equiv {}^{t+\Delta t}\mathbf{F} \cdot {}^{t}\mathbf{F}^{-1}, \quad {}^{\Delta t}\mathbf{R} \equiv {}^{t+\Delta t}\mathbf{R} \cdot {}^{t}\mathbf{R}^T. \tag{3.4}$$

The transposes of these tensors

$$^{\Delta t}\mathbf{F}^T = {}^{t}\mathbf{F}^{-T} \cdot {}^{t+\Delta t}\mathbf{F}^T, \quad {}^{\Delta t}\mathbf{R}^T = {}^{t}\mathbf{R} \cdot {}^{t+\Delta t}\mathbf{R}^T$$

are incrementally Lagrangian–Eulerian, and the tensors

$$^{\Delta t}\mathbf{F}^{-1} = {}^{t}\mathbf{F} \cdot {}^{t+\Delta t}\mathbf{F}^{-1}, \quad {}^{\Delta t}\mathbf{F}^{-T} = {}^{t+\Delta t}\mathbf{F}^{-T} \cdot {}^{t}\mathbf{F}^T$$

are incrementally Lagrangian–Eulerian and incrementally Eulerian–Lagrangian, respectively.

We introduce the notations

$$\Delta t_1 \equiv t_1 - t_0, \ \Delta t_2 \equiv t_2 - t_1, \ \ldots, \ \Delta t_N \equiv t_N - t_{N-1}.$$

The definition of the tensor $\mathbf{F}$ leads to the equality (see, e.g., [1])

$$^T\mathbf{F} = {}^{\Delta t_N}\mathbf{F} \cdot {}^{\Delta t_{N-1}}\mathbf{F} \cdot \ldots \cdot {}^{\Delta t_1}\mathbf{F}. \tag{3.5}$$

According to (3.3), we introduce the incremental rotation

$$^{\Delta t}\mathbf{Q} \equiv {}^{t+\Delta t}\mathbf{Q} \cdot {}^t\mathbf{Q}^T.$$

The following equality holds:

$$^T\mathbf{Q} = {}^{\Delta t_N}\mathbf{Q} \cdot {}^{\Delta t_{N-1}}\mathbf{Q} \cdot \ldots \cdot {}^{\Delta t_1}\mathbf{Q}. \tag{3.6}$$

Using equalities (3.5), (3.6) and the incremental objectivities of the tensors $^{\Delta t}\mathbf{F}$, $^{\Delta t}\mathbf{F}^T$, $^{\Delta t}\mathbf{F}^{-1}$, and $^{\Delta t}\mathbf{F}^{-T}$, we can confirm the absolute objectivities of the tensors $\mathbf{F}$, $\mathbf{F}^T$, $\mathbf{F}^{-1}$, and $\mathbf{F}^{-T}$ at discrete times $t_1, t_2, \ldots T$.

We introduce the *incremental right stretch tensor*

$$^{\Delta t}\mathbf{U} \equiv {}^{t+\Delta t}\mathbf{U} \cdot {}^t\mathbf{U}^{-1} \tag{3.7}$$

It can be shown that the tensor $^{\Delta t}\mathbf{U}$ is an absolutely Lagrangian tensor. We also introduce the *incremental left stretch tensor*

$$^{\Delta t}\tilde{\mathbf{V}} \equiv {}^{t+\Delta t}\mathbf{R} \cdot {}^{\Delta t}\mathbf{U} \cdot {}^{t+\Delta t}\mathbf{R}^T. \tag{3.8}$$

which is the absolutely Eulerian objective counterpart of the tensor $^{\Delta t}\mathbf{U}$. Note that the tensor $^{\Delta t}\tilde{\mathbf{V}}$ generally does not coincide with the tensor

$$^{\Delta t}\mathbf{V} \equiv {}^{t+\Delta t}\mathbf{V} \cdot {}^t\mathbf{V}^{-1}.$$

Equalities $(3.4)_2$, (3.7), and (3.8) lead to the following alternative expression for the tensor $^{\Delta t}\tilde{\mathbf{V}}$:

$$^{\Delta t}\tilde{\mathbf{V}} = {}^{t+\Delta t}\mathbf{V} \cdot {}^t\tilde{\mathbf{V}}^{-1}, \quad {}^t\tilde{\mathbf{V}} \equiv {}^{\Delta t}\mathbf{R} \cdot {}^t\mathbf{V} \cdot {}^{\Delta t}\mathbf{R}^T. \tag{3.9}$$

We choose some time interval $[t, \ t + \Delta t]$ in which we define the *relative deformation gradient tensor* $\mathbf{F}_r$ [2], choosing the configuration of the body $\mathfrak{B}$ at time t as the reference configuration and the configuration of this body at time $t + \Delta t$ as the current configuration. In this case, the following equality holds [2]:

$$\mathbf{F}_r = {}^{\Delta t}\mathbf{F}. \tag{3.10}$$

Applying the polar decomposition (2.1) to the tensor $\mathbf{F}_r$, we obtain

$$\mathbf{F}_r = \mathbf{R}_r \cdot \mathbf{U}_r = \mathbf{V}_r \cdot \mathbf{R}_r, \tag{3.11}$$

where the tensor $\mathbf{R}_r \in \mathcal{T}_{\mathrm{orth}}^{2+}$ is an incrementally Eulerian–Lagrangian *relative rotation* tensor and the tensors $\mathbf{U}_r$, $\mathbf{V}_r \in \mathcal{T}_{\mathrm{sym}}^2$ are incrementally Eulerian and Lagrangian *relative right and left stretch* positive definite tensors.

Note that the tensors $^{\Delta t}\mathbf{U}$ and $^{\Delta t}\tilde{\mathbf{V}}$ are absolutely objective tensors, and the tensors $\mathbf{U}_r$ and $\mathbf{V}_r$ are incrementally objective tensors. Note that the tensors $^{\Delta t}\mathbf{U}$ and $^{\Delta t}\tilde{\mathbf{V}}$ are generally nonsymmetric and the tensors $\mathbf{U}_r$ and $\mathbf{V}_r$ are symmetric positive definite tensors. The tensors $^{\Delta t}\mathbf{R}$ and $\mathbf{R}_r$ are both incremental Eulerian–Lagrangian tensors, but they are generally not equal, i.e., $^{\Delta t}\mathbf{R} \neq \mathbf{R}_r$. Next we find the relationship of the tensors $^{\Delta t}\mathbf{R}$, $^{\Delta t}\mathbf{U}$, and $^{\Delta t}\tilde{\mathbf{V}}$ to the tensors $\mathbf{R}_r$, $\mathbf{U}_r$, and $\mathbf{V}_r$.

From (2.1), (3.4), and (3.7), we obtain the equalities

$$^{\Delta t}\mathbf{F} = {}^{t+\Delta t}\mathbf{R} \cdot {}^{\Delta t}\mathbf{U} \cdot {}^{t}\mathbf{R}^T = {}^{t+\Delta t}\mathbf{V} \cdot {}^{\Delta t}\mathbf{R} \cdot {}^{t}\mathbf{V}^{-1}. \tag{3.12}$$

Using equalities (3.10), (3.11)$_1$, and (3.12)$_1$, we obtain

$$\mathbf{R}_r \cdot \mathbf{U}_r = {}^{t+\Delta t}\mathbf{R} \cdot {}^{\Delta t}\mathbf{U} \cdot {}^{t}\mathbf{R}^T. \tag{3.13}$$

Using equality (3.4)$_2$, we rewrite equality (3.13) as

$$\mathbf{R}_r \cdot \mathbf{U}_r = {}^{\Delta t}\mathbf{R} \cdot ({}^{t}\mathbf{R} \cdot {}^{\Delta t}\mathbf{U} \cdot {}^{t}\mathbf{R}^T). \tag{3.14}$$

Since the tensor in parentheses on the r.h.s. of (3.14) is generally nonsymmetric, it follows that generally $^{\Delta t}\mathbf{R} \neq \mathbf{R}_r$.

Equalities (3.10), (3.11)$_2$, and (3.12)$_2$ lead to the equality

$$\mathbf{V}_r \cdot \mathbf{R}_r = {}^{t+\Delta t}\mathbf{V} \cdot {}^{\Delta t}\mathbf{R} \cdot {}^{t}\mathbf{V}^{-1}. \tag{3.15}$$

After some transformations of the r.h.s. of (3.15), we rewrite this equality as

$$\mathbf{V}_r \cdot \mathbf{R}_r = {}^{\Delta t}\tilde{\mathbf{V}} \cdot {}^{\Delta t}\mathbf{R}. \tag{3.16}$$

Suppose that in the time interval $[t, t + \Delta t]$, the eigenindices of the tensors $\mathbf{U}$ and $\mathbf{V}$ do not change; then the following spectral representations of the tensors $^{t+\Delta t}\mathbf{U}$ and $^{t}\mathbf{U}^{-1}$ hold:

$$^{t+\Delta t}\mathbf{U} = \sum_{i=1}^{m} {}^{t+\Delta t}\lambda_i \, {}^{t+\Delta t}\mathbf{U}_i, \quad {}^{t}\mathbf{U}^{-1} = \sum_{i=1}^{m} {}^{t}\lambda_i^{-1} \, {}^{t}\mathbf{U}_i.$$

where $^{t+\Delta t}\lambda_i$ and $^{t}\lambda_i$ are the eigenvalues of the tensors $^{t+\Delta t}\mathbf{U}$ and $^{t}\mathbf{U}$, respectively, and $^{t+\Delta t}\mathbf{U}_i$ and $^{t}\mathbf{U}_i$ are the subordinate eigenprojections of these tensors ($i = 1, \ldots, m$). At $\Delta t \to 0$, sufficiently smooth motions have the limits $\lim_{\Delta t \to 0} {}^{t+\Delta t}\lambda_i = {}^{t}\lambda_i$ and $\lim_{\Delta t \to 0} {}^{t+\Delta t}\mathbf{U}_i = {}^{t}\mathbf{U}_i$ ($i = 1, \ldots, m$). This implies that for small Δt, the tensors $^{t+\Delta t}\mathbf{U}$ and $^{t}\mathbf{U}^{-1}$ are almost coaxial, whence it follows that the tensor $^{t+\Delta t}\mathbf{U} \cdot {}^{t}\mathbf{U}^{-1}$ is almost symmetric. Due to the uniqueness of the polar decomposition (3.11) (cf., [3]), it follows from (3.14) that for small Δt, the following equalities hold:

$$\mathbf{R}_r \approx {}^{\Delta t}\mathbf{R}, \quad \mathbf{U}_r \approx {}^{t}\mathbf{R} \cdot {}^{\Delta t}\mathbf{U} \cdot {}^{t}\mathbf{R}^{T}.$$

Then it follows from (3.16) that for small Δt,

$$\mathbf{V}_r \approx {}^{\Delta t}\tilde{\mathbf{V}}.$$

Note that although the incremental rotations ${}^{\Delta t}\mathbf{R}$ and $\mathbf{R}_r$ generally do not coincide, under IETs (3.1), they change according to the same law (3.2)$_3$; i.e., the following equalities hold:

$$
{}^{\Delta t}\mathbf{R}^{*} = {}^{t+\Delta t}\mathbf{Q} \cdot {}^{\Delta t}\mathbf{R} \cdot {}^{t}\mathbf{Q}^{T}, \quad \mathbf{R}_r^{*} = {}^{t+\Delta t}\mathbf{Q} \cdot \mathbf{R}_r \cdot {}^{t}\mathbf{Q}^{T}. \tag{3.17}
$$

Equalities (3.17) are a consequence of the incremental Eulerian–Lagrangian objectivity of the tensors ${}^{\Delta t}\mathbf{R}$ and $\mathbf{R}_r$.

References

1. M. Kleiber, P. Kowalczyk, *Introduction to Nonlinear Thermomechanics of Solids* (Springer, Switzerland, 2016)
2. J.C. Simo, T.J.R. Hughes, *Computational Inelasticity* (Springer, N.Y., 1998)
3. R.W. Ogden, *Non-linear Elastic Deformations* (Ellis Horwood, Chichester, 1984)

Chapter 4
Incrementally Objective Algorithms for Integrating CRs for Hooke-Like Hypoelastic Models in the Eulerian Form

Abstract This and the next chapters are central in this book. In Sect. 4.1, we recall the well-known definitions of generalized midpoint approximations, which are the basis for constructing both weak incrementally objective algorithms in the Hughes–Winget and Rubinstein–Atluri versions (Sect. 4.2) and strong incrementally objective algorithms (Sect. 4.3). It is noted that weak incrementally objective algorithms do not exactly reproduce SRBMs, whereas strong incrementally objective algorithms exactly reproduce them. In addition, the strong incrementally objective algorithms known in the literature are generalized in such a manner that they can be used to integrate CRs for hypoelastic material models based on kinematic variables dependent on the choice of the reference configuration. In Sect. 4.3, we analyze the previously published expressions for approximate incremental strain tensors and propose new expressions in the form of the Mooney incremental strain tensors, which have advantages over those available in the literature.

4.1 Generalized Midpoint Approximations

Suppose that we know the position vectors ${}^{t}\mathbf{x}$ and ${}^{t+\Delta t}\mathbf{x}$ of some material point of the body $\mathfrak{B}$ at discrete times t and $t + \Delta t$. We introduce a parameter $\alpha \in [0, 1]$ that interpolates the position vectors ${}^{\alpha}\mathbf{x}$ of the same material point for any time $\tau \in [t, t + \Delta t]$ according to the *generalized midpoint rule* (see, e.g., [1–4])

$$
{}^{\alpha}\mathbf{x} = (1 - \alpha){}^{t}\mathbf{x} + \alpha\, {}^{t+\Delta t}\mathbf{x}. \tag{4.1}
$$

From (4.1), the identification of notations holds

$$
{}^{0}\mathbf{x} = {}^{t}\mathbf{x}, \quad {}^{1}\mathbf{x} = {}^{t+\Delta t}\mathbf{x}. \tag{4.2}
$$

We introduce the following *relative deformation gradient tensors*:

$$
\mathbf{F}_r \equiv \frac{\partial\, {}^{1}\mathbf{x}}{\partial\, {}^{0}\mathbf{x}}, \quad {}^{\alpha}\mathbf{F}_r \equiv \frac{\partial\, {}^{\alpha}\mathbf{x}}{\partial\, {}^{0}\mathbf{x}}, \quad {}^{\alpha}\overset{+}{\mathbf{F}}_r \equiv \frac{\partial\, {}^{1}\mathbf{x}}{\partial\, {}^{\alpha}\mathbf{x}}. \tag{4.3}
$$

© The Author(s), under exclusive license to Springer Nature Switzerland AG 2023
S. Korobeynikov and A. Larichkin, *Objective Algorithms for Integrating Hypoelastic Constitutive Relations Based on Corotational Stress Rates*, SpringerBriefs in Continuum Mechanics, https://doi.org/10.1007/978-3-031-29632-1_4

Equality (3.10) still holds. From (4.3) we obtain

$$^{0}\mathbf{F}_r = \mathbf{I}, \quad ^{1}\mathbf{F}_r = \mathbf{F}_r, \quad ^{0}\overset{+}{\mathbf{F}}_r = \mathbf{F}_r, \quad ^{1}\overset{+}{\mathbf{F}}_r = \mathbf{I}. \tag{4.4}$$

Using the identity

$$\frac{\partial {}^{1}\mathbf{x}}{\partial {}^{0}\mathbf{x}} = \frac{\partial {}^{1}\mathbf{x}}{\partial {}^{\alpha}\mathbf{x}} \cdot \frac{\partial {}^{\alpha}\mathbf{x}}{\partial {}^{0}\mathbf{x}}$$

and notations (4.3), we obtain

$$\mathbf{F}_r = {}^{\alpha}\overset{+}{\mathbf{F}}_r \cdot {}^{\alpha}\mathbf{F}_r. \tag{4.5}$$

From (4.1)–(4.3), we obtain the equality

$$^{\alpha}\mathbf{F}_r = (1 - \alpha)\mathbf{I} + \alpha\mathbf{F}_r. \tag{4.6}$$

From (4.5) and (4.6), we have the equality

$$^{\alpha}\overset{+}{\mathbf{F}}_r = \mathbf{F}_r \cdot [(1 - \alpha)\mathbf{I} + \alpha\mathbf{F}_r]^{-1}.$$

We introduce the *incremental displacement vector*

$$\Delta\mathbf{u} \equiv {}^{t+\Delta t}\mathbf{x} - {}^{t}\mathbf{x} = {}^{1}\mathbf{x} - {}^{0}\mathbf{x}$$

and the *relative incremental displacement gradient tensor* (cf., [4])

$$^{\alpha}\mathbf{H} \equiv \frac{\partial \Delta\mathbf{u}}{\partial {}^{\alpha}\mathbf{x}}.$$

It can be shown that the following equalities hold:

$$^{\alpha}\mathbf{F}_r^{-1} = \mathbf{I} - \alpha{}^{\alpha}\mathbf{H}, \quad ^{\alpha}\overset{+}{\mathbf{F}}_r = \mathbf{I} + (1 - \alpha){}^{\alpha}\mathbf{H}. \tag{4.7}$$

The velocity gradient tensor can be represented in the form $(2.7)_2$. Since this tensor does not depend on the choice of the body reference configuration, we choose the configuration at time t as the reference configuration. Then equality $(2.7)_2$ is written as

$$\mathbf{l} = \dot{\mathbf{F}}_r \cdot \mathbf{F}_r^{-1}. \tag{4.8}$$

Consider expression (4.8) at time $t + \alpha\Delta t$

$$^{\alpha}\mathbf{l} = {}^{\alpha}\dot{\mathbf{F}}_r \cdot {}^{\alpha}\mathbf{F}_r^{-1}. \tag{4.9}$$

Using the generalized midpoint rule, we have [1]

$$^\alpha \dot{\mathbf{F}}_r \approx \frac{1}{\Delta t}(^1\mathbf{F}_r - {}^0\mathbf{F}_r) = \frac{1}{\Delta t}(\mathbf{F}_r - \mathbf{I}). \tag{4.10}$$

The latter equality was obtained using equalities $(4.4)_{1,2}$. Expression (4.10) approximates the tensor $^\alpha \dot{\mathbf{F}}_r$ with first-order accuracy for $\alpha = 0$ and $\alpha = 1$ and with second-order accuracy for $\alpha = 1/2$ (see, e.g., [5]).

We introduce the *approximate incremental deformation tensor* defined at time $t + \alpha \Delta t$

$$\Delta t^\alpha \mathbf{l} \equiv \text{approx}\left(\int_t^{t+\Delta t} \mathbf{l}\, d\tau \right). \tag{4.11}$$

Using equality (4.10) and definition (4.11), from (4.9) we obtain

$$\Delta t^\alpha \mathbf{l} = (\mathbf{F}_r - \mathbf{I}) \cdot {}^\alpha \mathbf{F}_r^{-1}. \tag{4.12}$$

It can be shown that the following equality holds:

$$^\alpha \mathbf{H} = (\mathbf{F}_r - \mathbf{I}) \cdot {}^\alpha \mathbf{F}_r^{-1}. \tag{4.13}$$

Using equality (4.13), we rewrite expression (4.12) as

$$\Delta t^\alpha \mathbf{l} = {}^\alpha \mathbf{H}. \tag{4.14}$$

We introduce the *approximate incremental strain* and the *approximate incremental vorticity tensors* defined at time $t + \alpha \Delta t$

$$\Delta t^\alpha \mathbf{d} \equiv \text{approx}\left(\int_t^{t+\Delta t} \mathbf{d}\, d\tau \right), \quad \Delta t^\alpha \mathbf{w} \equiv \text{approx}\left(\int_t^{t+\Delta t} \mathbf{w}\, d\tau \right). \tag{4.15}$$

We determine these tensors from the incremental counterparts of equalities $(2.8)_{2,3}$ using equality (4.14):

$$\Delta t^\alpha \mathbf{d} = \frac{1}{2}(^\alpha \mathbf{H} + {}^\alpha \mathbf{H}^T), \quad \Delta t^\alpha \mathbf{w} = \frac{1}{2}(^\alpha \mathbf{H} - {}^\alpha \mathbf{H}^T). \tag{4.16}$$

Note that these approximations of the incremental strain and vorticity tensors are used in [1].

4.2   Weak Incrementally Objective Algorithms

The development of incrementally objective algorithms for integrating CRs for the hypoelastic material model based on the Zaremba–Jaumann stress rate was pioneered by Hughes and Winget in the well-known paper [1], in which these authors developed the initial version of the weak algorithm using the second-order midpoint rule (see Sect. 4.2.1). Rubinstein and Atluri [6] improved this version by using an alternative approximation of the incremental rotation tensor (see Sect. 4.2.2). An extended family of objective algorithms was later developed by Simo and Hughes in [4] (see Sect. 4.2.3), who have shown that this family contains, in particular, the weak incrementally objective R-A algorithm. We show in this book (see Sect. 4.3) that some algorithms from the S-H family are not only weak but also strong incrementally objective.

4.2.1   Hughes–Winget Algorithm

Hughes and Winget [1] have found approximate proper orthogonal rotation tensor associated with the vorticity tensor $\mathbf{w}$ by approximately solving the tensorial equation (2.16) ($\boldsymbol{\omega} = \mathbf{w}$) in the time interval $[t, t + \Delta t]$. Denoting the increment of this tensor at time $t + \alpha \Delta t$ as $^\alpha \boldsymbol{\Psi}$ and applying the generalized midpoint rule, we have (see Eq. (13) in [1])

$$^\alpha \boldsymbol{\Psi} = \mathbf{I} + (\mathbf{I} - \alpha \, {}^\alpha \boldsymbol{\omega})^{-1} \cdot {}^\alpha \boldsymbol{\omega}. \tag{4.17}$$

Here for brevity, we introduced the notation $^\alpha \boldsymbol{\omega} \equiv \Delta t \, {}^\alpha \mathbf{w}$.

Definition 4.1 ([1]) Algorithms for integrating CRs for hypoelastic material models will be called *weak incrementally objective* if in the time interval $[t, t + \Delta t]$ for rigid motion (without strains) according to the law

$$^{t+\Delta t}\mathbf{x} = \tilde{\mathbf{Q}} \cdot {}^t \mathbf{x} \quad \forall \tilde{\mathbf{Q}} \in \mathcal{T}_{\text{orth}}^{2+} \tag{4.18}$$

the following equalities hold:

$$(1) \ \Delta t^\alpha \mathbf{d} = \mathbf{0}, \quad (2) \ {}^\alpha \boldsymbol{\Psi} = \tilde{\mathbf{Q}}.$$

Remark 4.1 In [1], weak incrementally objective algorithms are called *incrementally objective*. In our definition, we use the adjective *weak* according to [7–9]. Following these authors, we will differentiate between *weak* and *strong* incrementally objective algorithms.

In [3], approximations (4.16) were also obtained from the generalized midpoint rule. However, in that paper, CRs for a hypoelastic model based on the Oldroyd stress rate are integrated; therefore, the Cauchy problem (2.16) is not solved and the tensor

$^\alpha\boldsymbol{\Psi}$ is not determined. The definition of incremental objectivity (in our terminology, weak incremental objectivity) is formulated in Sect. 4.3 in [3] as follows:

Definition 4.2 ([3]) Algorithms for integrating CRs for hypoelastic material models will be called *weak incrementally objective* if

1. $\mathbf{F}_r \in \mathcal{T}^{2+}_{\text{orth}} \quad \Leftrightarrow \quad \Delta t^\alpha \mathbf{d} = \mathbf{0},$
2. $\mathbf{F}_r \in \mathcal{T}^{2+}_{\text{sym}} \quad \Leftrightarrow \quad \Delta t^\alpha \mathbf{w} = \mathbf{0}.$

Remark 4.2 The formulations of Definitions 4.1 and 4.2 are closely related to each other. Below, by default, by weak incrementally objective algorithms we mean algorithms according to Definition 4.1. References to weak incremental objectivity in the sense of Definition 4.2 will be made explicitly.

Hughes and Winget [1] have formulated and proved a theorem stating that approximations (4.16) and (4.17) ensure the weak incremental objectivity of the considered algorithm only for $\alpha = 1/2$. Pinsky et al. [3] have shown that the weak incremental objectivity of the algorithm in the sense of Definition 4.2 is achieved if and only if $\alpha = 1/2$ in approximations (4.16). Using equality (4.6), we can show that for $\alpha = 1/2$, expression (4.13) reduces to

$$^{1/2}\mathbf{H} = 2(\mathbf{F}_r - \mathbf{I}) \cdot (\mathbf{F}_r + \mathbf{I})^{-1}. \tag{4.19}$$

To simplify the expressions, we introduce the notations

$$\Delta t \mathbf{d}^m \equiv \Delta t^{1/2} \mathbf{d}, \quad \boldsymbol{\omega}^m \equiv \Delta t^{1/2} \mathbf{w}. \tag{4.20}$$

The tensors $\Delta t \mathbf{d}^m$ and $\boldsymbol{\omega}^m$ are expressed from (4.16) as

$$\Delta t \mathbf{d}^m = \frac{1}{2}(^{1/2}\mathbf{H} + {}^{1/2}\mathbf{H}^T), \quad \boldsymbol{\omega}^m = \frac{1}{2}(^{1/2}\mathbf{H} - {}^{1/2}\mathbf{H}^T), \tag{4.21}$$

where the tensor $^{1/2}\mathbf{H}$ is defined in (4.19).

We introduce the notations

$$^{\Delta t}\boldsymbol{\Psi} \equiv {}^{1/2}\boldsymbol{\Psi}, \quad {}^{\Delta t/2}\boldsymbol{\Psi} \equiv \sqrt{^{1/2}\boldsymbol{\Psi}}. \tag{4.22}$$

In view of notations (4.21)$_2$ and (4.22)$_1$, expression (4.17) for $\alpha = 1/2$ reduces to the following form (cf., [1]):

$$^{\Delta t}\boldsymbol{\Psi} = \left(\mathbf{I} + \frac{1}{2}\boldsymbol{\omega}^m\right) \cdot \left(\mathbf{I} - \frac{1}{2}\boldsymbol{\omega}^m\right)^{-1}. \tag{4.23}$$

A more difficult problem is to determine the tensor $^{\Delta t/2}\boldsymbol{\Psi}$, especially in the case of 3D analysis (cf., [10, 11]). Nevertheless, in 2D analysis, it is fairly easy to determine the tensor $^{\Delta t/2}\boldsymbol{\Psi}$ using the algorithm presented in [12] (see also [11, 13]).

Specifying expression $(2.38)_2$ for the tensor $\boldsymbol{\tau}$ for the integration of Eq. (2.43) using the midpoint rule at $\alpha = 1/2$, we obtain the following expression for the tensor $^{t+\Delta t}\boldsymbol{\tau}$ (cf., [12], see also [11]) for the Hughes–Winget algorithm (the *H–W algorithm*):

$$^{t+\Delta t}\boldsymbol{\tau} = {}^{\Delta t}\boldsymbol{\Psi} \cdot {}^{t}\boldsymbol{\tau} \cdot {}^{\Delta t}\boldsymbol{\Psi}^{T} + 2\mu{}^{\Delta t/2}\boldsymbol{\Psi} \cdot (\Delta t\mathbf{d}^{m}) \cdot {}^{\Delta t/2}\boldsymbol{\Psi}^{T}, \qquad (4.24)$$

where the tensor $\Delta t\mathbf{d}^{m}$ is defined in $(4.21)_1$ and the tensor $^{\Delta t}\boldsymbol{\Psi}$ in (4.23).

Remark 4.3 Restrictions on the time step Δt in the H–W algorithm follow from the form of the second multiplier on the r.h.s. of (4.19). In particular, the expression on the r.h.s. of (4.19) loses its meaning if $\det(\mathbf{F}_r + \mathbf{I}) = 0$, which can occur, e.g., for 2D rotation of the body for a rotation angle increment of π (cf., [1]). In this (critical) situation, it is necessary, e.g., to repeat the calculations with reduced time steps.

Remark 4.4 In [1], the stress tensor $^{t+\Delta t}\boldsymbol{\tau}$ is updated using an expression similar to (4.24) in which the tensor $^{\Delta t/2}\boldsymbol{\Psi}$ is replaced by the tensor $\mathbf{I}$. Expression (4.24) strictly corresponding to the midpoint rule was obtained by Key and Krieg [12].

4.2.2 Rubinstein–Atluri Algorithm and Its Generalization

The Rubinstein–Atluri algorithm (the *R–A algorithm*) [6], like the H–W algorithm, uses expressions (4.21) to approximate incremental strains and vorticity tensors. However, in this algorithm, the incremental rotation tensor $^{\Delta t}\boldsymbol{\Psi}$ is determined using the Rodrigues formula (see, e.g., [6, 14] and Appendix C) instead of expression (4.23). For 2D analysis, this formula is written as follows (see, e.g., [6, 11, 15] and Appendix C):

$$^{\Delta t}\boldsymbol{\Psi} = \begin{bmatrix} \cos\omega_{12}^{m} & \sin\omega_{12}^{m} \\ -\sin\omega_{12}^{m} & \cos\omega_{12}^{m} \end{bmatrix} = \begin{bmatrix} \cos\omega_{21}^{m} & -\sin\omega_{21}^{m} \\ \sin\omega_{21}^{m} & \cos\omega_{21}^{m} \end{bmatrix}. \qquad (4.25)$$

Here ω_{12}^{m} and ω_{21}^{m} ($\omega_{21}^{m} = -\omega_{12}^{m}$) are the components of the tensor $\boldsymbol{\omega}^{m}$ in the Cartesian coordinate system

$$\boldsymbol{\omega}^{m} = \begin{bmatrix} 0 & \omega_{12}^{m} \\ \omega_{21}^{m} & 0 \end{bmatrix}.$$

It can be shown (cf., [16]) that expression (4.23) approximates expression (4.25).

To determine the tensor $^{t+\Delta t}\boldsymbol{\tau}$, the R–A algorithm uses expression (4.24). This algorithm, like the H–W algorithm, is weak incrementally objective. Unlike the H–W algorithm, the R–A algorithm determines the tensor $^{\Delta t/2}\boldsymbol{\Psi}$ from expression (4.25) by replacing the quantity ω_{12}^{m} by the quantity $\omega_{12}^{m}/2$. For the R–A algorithm, as for the H–W algorithm, a degeneration of the tensor $\mathbf{F}_r + \mathbf{I}$ is possible; in this case, the expression on the r.h.s. of (4.19) loses its meaning. In this situation, one can, e.g., change the integration time step (see Remark 4.3).

In the original form, the R–A algorithm was developed to integrate CRs for hypoe-lastic material models based on the Zaremba–Jaumann stress rate. We generalize this algorithm for G-models using the incremental counterpart of expression $(2.46)_2$ and rewriting the expression for ω^m (see $(4.20)_2$) as follows:

$$\omega^m = \Delta t^{1/2} \mathbf{w} + \sum_{i \neq j = 1}^{m} {}^{t+\Delta t/2}g_{ij} {}^{t+\Delta t/2}\mathbf{b}_i \cdot (\Delta t \mathbf{d}^m) \cdot {}^{t+\Delta t/2}\mathbf{b}_j. \tag{4.26}$$

Here the tensor $\Delta t \mathbf{d}^m$ is determined from expression $(4.21)_1$ and the tensor $\Delta t^{1/2}\mathbf{w}$ is determined from expression $(4.21)_2$ by replacing the tensor ω^m on the l.h.s. of this equality by the tensor $\Delta t^{1/2}\mathbf{w}$, i.e.,

$$\Delta t^{1/2}\mathbf{w} = \frac{1}{2}({}^{1/2}\mathbf{H} - {}^{1/2}\mathbf{H}^T). \tag{4.27}$$

The quantities ${}^{t+\Delta t/2}g_{ij}$ (see Eq. (2.47)) and ${}^{t+\Delta t/2}\mathbf{b}_i (= {}^{t+\Delta t/2}\mathbf{V}_i)$, ${}^{t+\Delta t/2}\mathbf{b}_j (= {}^{t+\Delta t/2}\mathbf{V}_j)$ $(i, j = 1, \ldots, m)$ are determined from the spectral representation of the left Piola deformation tensor $\mathbf{b}$ at time $t + \Delta t/2$. The remaining expressions for determining the tensors ${}^{\Delta t}\boldsymbol{\Psi}$, ${}^{\Delta t/2}\boldsymbol{\Psi}$, and ${}^{t+\Delta t}\boldsymbol{\tau}$ remain the same.

Remark 4.5 The weak incrementally objective algorithm presented in this section can be modified to integrate CRs for the hypoelastic model based on the Green–Naghdi stress rate. In this case, instead of determining the tensor ω^m from expressions (4.26) and (4.27) and then using expression (4.25) to determine the tensors ${}^{\Delta t}\boldsymbol{\Psi}$ and ${}^{\Delta t/2}\boldsymbol{\Psi}$, the latter tensors can be determined from the expressions

$$\Delta t \mathbf{R} (= {}^{\Delta t}\boldsymbol{\Psi}) = {}^{t+\Delta t}\mathbf{R} \cdot {}^t\mathbf{R}^T, \quad {}^{\Delta t/2}\mathbf{R} (= {}^{\Delta t/2}\boldsymbol{\Psi}) = {}^{t+\Delta t}\mathbf{R} \cdot {}^{t+\Delta t/2}\mathbf{R}^T, \tag{4.28}$$

where the tensors ${}^{t+\Delta t}\mathbf{R}$, ${}^{t+\Delta t/2}\mathbf{R}$, and ${}^t\mathbf{R}$ are determined from the tensors ${}^{t+\Delta t}\mathbf{F}$, ${}^{t+\Delta t/2}\mathbf{F}$, and ${}^t\mathbf{F}$, respectively, using their polar decompositions (see Appendix B). In our classification of algorithms, this approach to determining updated stresses is mixed since in this approach, the approximate incremental strain tensor $\Delta t \mathbf{d}^m$ is defined from $(4.21)_1$ due to weak incrementally objective algorithm, but the tensors ${}^{\Delta t}\boldsymbol{\Psi}$ and ${}^{\Delta t/2}\boldsymbol{\Psi}$ are defined from (4.28) due to absolutely objective algorithm (see Chap. 5). Similar mixed algorithms for integrating CRs for the hypoelastic model based on the Green–Naghdi stress rate are proposed in [3, 13, 16–18]. Nevertheless, Flanagan and Taylor [19] argue that for the integration of such CRs implemented in any explicit FE codes, the determination of the rotation tensor from the polar decomposition of the deformation gradient tensor is too expensive, and they, in fact, propose to determine the incremental rotation tensor ${}^{\Delta t}\mathbf{R}$ from expressions similar to (4.26) and (4.27) with subsequent use of the H–W algorithm.

Remark 4.6 In the absence of initial stresses (i.e., $\sigma^0 = 0$ at $t = t_0$), the hypoelastic model based on the logarithmic stress rate is the rate form of the Hencky isotropic hyperelastic model (cf., [20–22], see also [23–25]). In this case, there is no need to integrate CRs (2.39) or (2.40) since these hypoelastic CRs are equivalent to the hyperelastic CRs

$$\bar{\tau} = \lambda \mathrm{tr} \mathbf{E}^{(0)} \, \mathbf{I} + 2\mu \mathbf{E}^{(0)}, \quad \tau = \lambda \mathrm{tr} \mathbf{e}^{(0)} \, \mathbf{I} + 2\mu \mathbf{e}^{(0)}$$

in the Lagrangian and Eulerian forms, respectively (see, e.g., [25–27]). However, in the presence of initial stresses (i.e., $\sigma^0 \neq 0$ at $t = t_0$), this hypoelasticity model does not have a counterpart in the form of any hyperelastic model (cf., [28]). In this case, CRs for the hypoelastic model based on the logarithmic stress rate can be integrated using the expressions given above (see Eqs. (4.24)–(4.27)).

4.2.3 Simo–Hughes Algorithm and Its Generalization

In the algorithm proposed by Simo and Hughes (cf., [4]) (the *S–H algorithm*), the approximate incremental strain tensor $\Delta t \mathbf{d}^a$ is found from equality (2.19)$_2$. Using the generalized midpoint rule, we obtain the following expression for the approximate incremental strain tensor (cf., Eq. (8.1.13) in [4]):

$$\Delta t \, {}^\alpha \mathbf{d}_1 = {}^\alpha \mathbf{F}_r^{-T} \cdot \mathbf{E}_r^{(2)} \cdot {}^\alpha \mathbf{F}_r^{-1}, \quad \mathbf{E}_r^{(2)}(\equiv {}^1 \mathbf{E}_r^{(2)}) \equiv \frac{1}{2}(\mathbf{F}_r^T \cdot \mathbf{F}_r - \mathbf{I}). \tag{4.29}$$

Here we introduced the (incrementally Lagrangian-objective) *incremental Green–Lagrange strain tensor* $\mathbf{E}_r^{(2)}$ (see Table 2.1).

Using equality (4.5), we obtain the following expression for the tensor $\Delta t \, {}^\alpha \mathbf{d}_1$, which is alternative to (4.29)$_1$ (cf., Eq. (8.1.31) in [4]):

$$\Delta t \, {}^\alpha \mathbf{d}_1 = {}^\alpha \overset{+}{\mathbf{F}}_r^T \cdot \mathbf{e}_r^{(-2)} \cdot {}^\alpha \overset{+}{\mathbf{F}}_r. \tag{4.30}$$

Here the tensor $\mathbf{e}_r^{(-2)}$ is the (incrementally Eulerian-objective) *incremental Almansi strain tensor* (see Table 2.1)

$$\mathbf{e}_r^{(-2)} \equiv \frac{1}{2}(\mathbf{I} - \mathbf{F}_r^{-T} \cdot \mathbf{F}_r^{-1}). \tag{4.31}$$

To determine the approximate incremental strain tensor, Simo and Hughes used a representation of the stretching tensor $\mathbf{d}$ in the form of the lower Oldroyd rate of the Almansi strain tensor $\mathbf{e}^{(-2)}$. It has been shown [24, 29] that for the stretching tensor $\mathbf{d}$, one can find infinitely many representations in the form of some objective rates of strain tensors from the Hill family. We use a representation of the stretching tensor $\mathbf{d}$ in the form of the upper Oldroyd rate of the Finger strain tensor $\mathbf{e}^{(2)}$ (see (2.19)$_1$)

to generalize the S–H algorithm. Using the generalized midpoint rule, we obtain the following expression for the approximate incremental strain tensor:

$$\Delta t\,{}^{\alpha}\mathbf{d}_2 = {}^{\alpha}\mathbf{F}_r \cdot \mathbf{E}_r^{(-2)} \cdot {}^{\alpha}\mathbf{F}_r^{T}, \quad \mathbf{E}_r^{(-2)}(\equiv {}^{1}\mathbf{E}_r^{(-2)}) \equiv \frac{1}{2}(\mathbf{I} - \mathbf{F}_r^{-1} \cdot \mathbf{F}_r^{-T}). \quad (4.32)$$

Here we introduced the (incrementally Lagrangian-objective) *incremental Karni–Reiner strain tensor* $\mathbf{E}_r^{(-2)}$ (see Table 2.1).

Using equality (4.5), we obtain the following expression for the tensor $\Delta t\,{}^{\alpha}\mathbf{d}_2$, which is alternative to $(4.32)_1$:

$$\Delta t\,{}^{\alpha}\mathbf{d}_2 = {}^{\alpha}\overset{+}{\mathbf{F}}_r^{-1} \cdot \mathbf{e}_r^{(2)} \cdot {}^{\alpha}\overset{+}{\mathbf{F}}_r^{-T}. \quad (4.33)$$

Here the tensor $\mathbf{e}_r^{(2)}$ is the (incrementally Eulerian-objective) *incremental Finger strain tensor* (see Table 2.1):

$$\mathbf{e}_r^{(2)} \equiv \frac{1}{2}(\mathbf{F}_r \cdot \mathbf{F}_r^{T} - \mathbf{I}). \quad (4.34)$$

We now express the tensors $\Delta t\,{}^{\alpha}\mathbf{d}_1$ and $\Delta t\,{}^{\alpha}\mathbf{d}_2$ in terms of the tensor ${}^{\alpha}\mathbf{H}$. Substituting expression $(4.29)_2$ into equality $(4.29)_1$ and using equality (4.5), we express the tensor $\Delta t\,{}^{\alpha}\mathbf{d}_1$ as

$$\Delta t\,{}^{\alpha}\mathbf{d}_1 = \frac{1}{2}({}^{\alpha}\overset{+}{\mathbf{F}}_r^{T} \cdot {}^{\alpha}\overset{+}{\mathbf{F}}_r - {}^{\alpha}\mathbf{F}_r^{-T} \cdot {}^{\alpha}\mathbf{F}_r^{-1}). \quad (4.35)$$

Using equalities (4.7), from (4.35) we obtain the final expression for the tensor $\Delta t\,{}^{\alpha}\mathbf{d}_1$ (cf., Eq. (8.1.22) in [4]):

$$\Delta t\,{}^{\alpha}\mathbf{d}_1 = \frac{1}{2}[{}^{\alpha}\mathbf{H} + {}^{\alpha}\mathbf{H}^{T} + (1 - 2\alpha){}^{\alpha}\mathbf{H}^{T} \cdot {}^{\alpha}\mathbf{H}]. \quad (4.36)$$

Substituting expression $(4.32)_2$ into equality $(4.32)_1$ and using equality (4.5), we express the tensor $\Delta t\,{}^{\alpha}\mathbf{d}_2$ as

$$\Delta t\,{}^{\alpha}\mathbf{d}_2 = \frac{1}{2}({}^{\alpha}\mathbf{F}_r \cdot {}^{\alpha}\mathbf{F}_r^{T} - {}^{\alpha}\overset{+}{\mathbf{F}}_r^{-1} \cdot {}^{\alpha}\overset{+}{\mathbf{F}}_r^{-T}). \quad (4.37)$$

Using equalities (4.7), from (4.37) we obtain the following expression for the tensor $\Delta t\,{}^{\alpha}\mathbf{d}_2$:

$$\Delta t\,{}^{\alpha}\mathbf{d}_2 = \frac{1}{2}\{(\mathbf{I} - \alpha{}^{\alpha}\mathbf{H})^{-1} \cdot (\mathbf{I} - \alpha{}^{\alpha}\mathbf{H}^{T})^{-1} - [\mathbf{I} + (1 - \alpha){}^{\alpha}\mathbf{H}]^{-1} \cdot [\mathbf{I} + (1 - \alpha){}^{\alpha}\mathbf{H}^{T}]^{-1}\}. \quad (4.38)$$

We represent the expression on the r.h.s. of (4.38) using the Taylor formula. Keeping terms up to the second order in ${}^{\alpha}\mathbf{H}$ and performing lengthy calculations, we obtain the following approximate expression for the tensor $\Delta t\,{}^{\alpha}\mathbf{d}_2$:

$$\Delta t\,{}^{\alpha}\mathbf{d}_2 \approx \frac{1}{2}\{{}^{\alpha}\mathbf{H} + {}^{\alpha}\mathbf{H}^T + (2\alpha - 1)[({}^{\alpha}\mathbf{H})^2 + ({}^{\alpha}\mathbf{H}^T)^2 + {}^{\alpha}\mathbf{H} \cdot {}^{\alpha}\mathbf{H}^T]\}. \tag{4.39}$$

To establish the weak incremental objectivity of the S–H algorithm, we assume that in the time interval $[t, t + \Delta t]$, the law of motion is given by (4.18), whence we obtain

$$\mathbf{F}_r = \mathbf{R}_r = \tilde{\mathbf{Q}} \in \mathcal{T}_{\text{orth}}^{2+}.$$

Then $(4.29)_2$ and $(4.32)_2$ imply that $\mathbf{E}_r^{(2)} = \mathbf{E}_r^{(-2)} = \mathbf{0}$, and $(4.29)_1$ and $(4.32)_1$ lead to the equalities

$$\Delta t\,{}^{\alpha}\mathbf{d}_1 = \Delta t\,{}^{\alpha}\mathbf{d}_2 = \mathbf{0} \quad \forall\, \alpha \in [0, 1].$$

In the S–H algorithm, the tensor ${}^{\Delta t}\boldsymbol{\Psi}$ is determined using the Rodrigues formula, which for 2D analysis is given by (4.25), where the tensor $\boldsymbol{\omega}^m$ is determined from expression $(4.16)_2$ taking into account expression (4.13). In this case, the tensor ${}^{\Delta t}\boldsymbol{\Psi}$ is proper orthogonal for any value of $\alpha \in [0, 1]$ (see Remark 8.3.3, item #1 in [4]). Thus, Definition 4.1 implies that the S–H algorithm for determining the tensors $\Delta t\,{}^{\alpha}\mathbf{d}_1$, $\Delta t\,{}^{\alpha}\mathbf{d}_2$, and ${}^{\Delta t}\boldsymbol{\Psi}$ is weak incrementally objective $\forall\, \alpha \in [0, 1]$. For $\alpha = 1/2$, this algorithm is second-order accurate in time, and for $\alpha = 0, 1$, it is first-order accurate in time. Note also that $(4.21)_1$, (4.36), and (4.39) lead to the equalities

$$\Delta t\,{}^{1/2}\mathbf{d}_1 = \Delta t\,{}^{1/2}\mathbf{d}_2 = \Delta t\,\mathbf{d}^m.$$

Thus, the S–H algorithm for determining the tensors $\Delta t\,\mathbf{d}^m$ and ${}^{\Delta t}\boldsymbol{\Psi}$ for $\alpha = 1/2$ reduces to the R–A algorithm, and in this case, the tensor ${}^{t+\Delta t}\boldsymbol{\tau}$ is determined from expression (4.24). For $\alpha \neq 1/2, 0, 1$, this algorithm has not found application; however, for $\alpha = 0, 1$ the S–H algorithm is not only weak incrementally objective, but also strong incrementally objective (see Sect. 4.3).

Note that flowchart of weak incrementally objective algorithms is given in Fig. A.2.

4.3 Strong Incrementally Objective Algorithms

Reed and Atluri [30, 31] have noted that the requirement of weak incremental objectivity for approximations of physical quantities may be insufficient for a body at large strains with large SRBMs. They introduced the definition of *absolute objectivity* both for approximations of physical quantities and for algorithms implementing these approximations. Absolutely objective algorithms for integrating CRs for hypoelastic material models will be developed in Chap. 5 of this book. Note that strong incrementally objective algorithms for integrating CRs for hypoelastic models will be developed for the Eulerian formulations of these CRs presented in Sect. 2.3. The ambiguity in the definition of approximate incremental strain and rotation tensors has led to different versions of strong incrementally objective algorithms introduced previously [5, 7–9, 26, 32–41]. In this section, we analyze these algorithms and propose new strong incrementally objective algorithms that improve the algorithms proposed in the above papers and can also be used to integrate CRs for hypoelastic material models that are not considered in the above literature.

4.3.1   Necessary Conditions for the Strong Incremental Objectivity of the S–H Algorithm

Theorem 4.1 *A necessary condition for the strong incremental objectivity of the S–H algorithm is the coincidence of the point $^\alpha\mathbf{x}$ with one of the edge points of the time interval $[t, t + \Delta t]$; i.e., in this case, the parameter α can take only two values: $\alpha = 0$ ($^0\mathbf{x} = {}^t\mathbf{x}$) or $\alpha = 1$ ($^1\mathbf{x} = {}^{t+\Delta t}\mathbf{x}$).*

Proof In the time interval $[t, t + \Delta t]$, the basic kinematic quantity is the relative deformation gradient $\mathbf{F}_r$. The midpoint time $t + \alpha\Delta t$ divides the time interval $[t, t + \Delta t]$ into two subintervals $[t, t + \alpha\Delta t]$ and $[t + \alpha\Delta t, t + \Delta t]$ in such a manner that $[t, t + \Delta t] = [t, t + \alpha\Delta t] \cup [t + \alpha\Delta t, t + \Delta t]$. Equalities (3.5) and (4.5) imply that the tensors $^\alpha\mathbf{F}_r$ and $^\alpha\overset{+}{\mathbf{F}}_r$ have the meaning of the relative deformation gradients for the subintervals $[t, t + \alpha\Delta t]$ and $[t + \alpha\Delta t, t + \Delta t]$, respectively. If the point $^\alpha\mathbf{x}$ defined by (4.1) is a "legitimate" point in the law of motion $\mathbf{x} = \mathbf{x}(\mathbf{X}, t)$, then, under the IETs (3.1), the relative deformation gradients $^\alpha\mathbf{F}_r$ and $^\alpha\overset{+}{\mathbf{F}}_r$ should satisfy the incremental Eulerian–Lagrangian objectivity conditions $(3.2)_3$:

$$(^\alpha\mathbf{F}_r)^* = {}^{t+\alpha\Delta t}\mathbf{Q} \cdot {}^\alpha\mathbf{F}_r \cdot {}^t\mathbf{Q}^T, \tag{4.40}$$

$$(^\alpha\overset{+}{\mathbf{F}}_r)^* = {}^{t+\Delta t}\mathbf{Q} \cdot {}^\alpha\overset{+}{\mathbf{F}}_r \cdot {}^{t+\alpha\Delta t}\mathbf{Q}^T.$$

We now determine the values of the parameter $\alpha \in [0, 1]$ for which equalities (4.40) hold for any tensors $^t\mathbf{Q}$, $^{t+\alpha\Delta t}\mathbf{Q}$, and $^{t+\Delta t}\mathbf{Q}$.

Let equality $(4.40)_1$ hold. Using the incremental Eulerian–Lagrangian objectivity of the tensor $\mathbf{F}_r$, from (4.6) we find that under the IETs (3.1), the tensor $^\alpha\mathbf{F}_r$ changes as

$$(^\alpha\mathbf{F}_r)^* = [(1 - \alpha){}^t\mathbf{Q} + \alpha{}^{t+\Delta t}\mathbf{Q} \cdot \mathbf{F}_r] \cdot {}^t\mathbf{Q}^T. \tag{4.41}$$

On the other hand, substitution of the expression for $^\alpha\mathbf{F}_r$ from (4.6) into the r.h.s. of $(4.40)_1$ yields

$$(^\alpha\mathbf{F}_r)^* = {}^{t+\alpha\Delta t}\mathbf{Q} \cdot [(1 - \alpha)\mathbf{I} + \alpha\mathbf{F}_r] \cdot {}^t\mathbf{Q}^T. \tag{4.42}$$

Equating the expressions on r.h.s. of (4.41) and (4.42) and taking into account the arbitrariness of the tensor $^t\mathbf{Q}$, we arrive at the equality

$$^{t+\alpha\Delta t}\mathbf{Q} = [(1 - \alpha){}^t\mathbf{Q} + \alpha{}^{t+\Delta t}\mathbf{Q} \cdot \mathbf{F}_r] \cdot [(1 - \alpha)\mathbf{I} + \alpha\mathbf{F}_r]^{-1} \tag{4.43}$$

A necessary condition for the tensor $^{t+\alpha\Delta t}\mathbf{Q}$ to be a proper orthogonal tensor is the equality

$$^{t+\alpha\Delta t}\mathbf{Q}^{-1} = {}^{t+\alpha\Delta t}\mathbf{Q}^T . \tag{4.44}$$

After elementary transformations, from (4.43) and (4.44) we obtain the necessary condition for the proper orthogonality of the tensor $^{t+\alpha\Delta t}\mathbf{Q}$:

$$\alpha(1 - \alpha)(\mathbf{F}_r + \mathbf{F}_r^T) = \alpha(1 - \alpha)({}^t\mathbf{Q}^T \cdot {}^{t+\Delta t}\mathbf{Q} \cdot \mathbf{F}_r + \mathbf{F}_r^T \cdot {}^{t+\Delta t}\mathbf{Q}^T \cdot {}^t\mathbf{Q}). \tag{4.45}$$

For arbitrary tensors $\mathbf{F}_r$, equality (4.45) reduces to the equalities

$$(1)\ \alpha = 0; \quad (2)\ \alpha = 1; \quad (3)\ {}^t\mathbf{Q} = {}^{t+\Delta t}\mathbf{Q}. \tag{4.46}$$

Since for an arbitrary tensor $\mathbf{Q}(t)$, equality $(4.46)_3$ cannot be satisfied, the necessary condition for the proper orthogonality of the tensor $^{t+\alpha\Delta t}\mathbf{Q}$ reduces to equalities $(4.46)_{1,2}$. Using equalities $(4.4)_{3,4}$ and the fact that the tensor $\mathbf{F}_r$ is Eulerian–Lagrangian, it is easy to check that for $\alpha = 0, 1$, equality $(4.40)_2$ is identically satisfied. □

Corollary 4.1 *The S–H algorithm based on second-order accurate midpoint approximations ($\alpha = 1/2$) is not strong incrementally objective. Only the S–H algorithm based on first-order accurate generalized midpoint approximations for $\alpha = 0$ and $\alpha = 1$ can be strong incrementally objective.*

4.3.2 Expressions for Approximate Incremental Strains for Strong Incrementally Objective Algorithms

Theorem 4.1 implies that the tensors $\Delta t\,{}^0\mathbf{d}_1$, $\Delta t\,{}^1\mathbf{d}_1$, $\Delta t\,{}^0\mathbf{d}_2$, and $\Delta t\,{}^1\mathbf{d}_2$ obtained in Sect. 4.2.3 satisfy the necessary conditions of strong incremental objectivity for the S–H algorithm. We now check whether these tensors satisfy the incremental objectivity conditions formulated in Chap. 3. Using equalities (4.4), (4.29), (4.30), (4.32), and (4.33), we obtain the following approximate incremental strains:

$$\Delta t\,{}^0\mathbf{d}_1 = \mathbf{E}_r^{(2)}, \quad \Delta t\,{}^1\mathbf{d}_1 = \mathbf{e}_r^{(-2)}, \quad \Delta t\,{}^0\mathbf{d}_2 = \mathbf{E}_r^{(-2)}, \quad \Delta t\,{}^1\mathbf{d}_2 = \mathbf{e}_r^{(2)}.$$

Next, to unify the notations, we introduce a new notations as follows:

$$\Delta t\bar{\mathbf{d}}_1 \equiv \Delta t\,{}^0\mathbf{d}_1 = \mathbf{E}_r^{(2)}, \quad \Delta t\mathbf{d}_1 \equiv \Delta t\,{}^1\mathbf{d}_2 = \mathbf{e}_r^{(2)},$$
$$\Delta t\bar{\mathbf{d}}_2 \equiv \Delta t\,{}^0\mathbf{d}_2 = \mathbf{E}_r^{(-2)}, \quad \Delta t\mathbf{d}_2 \equiv \Delta t\,{}^1\mathbf{d}_1 = \mathbf{e}_r^{(-2)}.$$

Using the incremental Eulerian–Lagrangian objectivity of the tensor $\mathbf{F}_r$ (see Chap. 3) and definitions $(4.29)_2$, (4.31), $(4.32)_2$, and (4.34) of the tensors $\mathbf{E}_r^{(2)}$, $\mathbf{e}_r^{(-2)}$, $\mathbf{E}_r^{(-2)}$, and $\mathbf{e}_r^{(2)}$, respectively, we establish according to Definition 3.1 that the tensors $\Delta t \bar{\mathbf{d}}_1$ and $\Delta t \bar{\mathbf{d}}_2$ are incrementally Lagrangian-objective and the tensors $\Delta t \mathbf{d}_1$ and $\Delta t \mathbf{d}_2$ are incrementally Eulerian-objective. Thus, the tensors $\Delta t \bar{\mathbf{d}}_1$, $\Delta t \bar{\mathbf{d}}_2$, $\Delta t \mathbf{d}_1$, and $\Delta t \mathbf{d}_2$ can be approximate incremental strain tensors for strong incrementally objective algorithms.

Remark 4.7 Incrementally Eulerian-objective expressions were also obtained for the approximate incremental strain $\Delta t \mathbf{d}_1$ in [33–35] and for $\Delta t \mathbf{d}_2$ in [32].

Using equality $(2.19)_3$, we obtain one more approximate incremental strain tensor. Rewrite this equality as follows ($\ln \mathbf{V} \equiv \mathbf{e}^{(0)}$)

$$\overline{\boldsymbol{\Psi}_{\log}^{ET} \cdot \ln \mathbf{V} \cdot \boldsymbol{\Psi}_{\log}^{E}} = \boldsymbol{\Psi}_{\log}^{ET} \cdot \mathbf{d} \cdot \boldsymbol{\Psi}_{\log}^{E}. \tag{4.47}$$

We fit together the reference configuration of the body $\mathfrak{B}$ and its configuration at time t. Using the backward finite difference scheme (the generalized midpoint rule with $\alpha = 1$) and taking into account the equalities $\ln \mathbf{V} = 0$ at time t and $\ln \mathbf{V} = \ln \mathbf{V}_r$ at time $t + \Delta t$, from (4.47) we obtain the approximate incremental strain

$$\Delta t \mathbf{d}_3 \equiv \ln \mathbf{V}_r. \tag{4.48}$$

To obtain the incremental rotation tensor $^{\Delta t} \boldsymbol{\Psi}_{\log}^{E}$, it is necessary, in the time interval $[t, t + \Delta t]$, to solve Eq. (2.16), where the tensor $\boldsymbol{\omega} = \boldsymbol{\omega}^{\log}$ is determined from the incremental counterpart of Eq. $(2.44)_2$. It follows from expression $(2.13)_2$ that expression (4.48) approximates only the component of the tensor $\mathbf{d}$ that is coaxial with the tensor $\mathbf{V}$ and ignores the component of the tensor $\mathbf{d}$ that is orthogonal to the tensor $\mathbf{V}$. Taking λ_i and $\mathbf{V}_i$ ($i = 1, \ldots, m$) on the r.h.s. of $(2.44)_2$ to be the eigenvalues and subordinate eigenprojections of the tensor $\mathbf{V}_r$, we see that, in view of (4.48), the second term on the r.h.s. of $(2.44)_2$ is equal to the zero tensor, whence it follows that the expression for the approximate incremental strain (4.48) is compatible with the following expression for the tensor $^{\Delta t} \boldsymbol{\Psi}_{\log}^{E}$:

$$^{\Delta t} \boldsymbol{\Psi}_{\log}^{E} = \mathbf{R}_r. \tag{4.49}$$

Using the forward finite difference scheme (the generalized midpoint rule with $\alpha = 0$), from (4.47) and (4.49), we obtain the following approximate incremental strain:

$$\Delta t \bar{\mathbf{d}}_3 \equiv \ln \mathbf{U}_r.$$

It is easy to show that the tensor $\Delta t \bar{\mathbf{d}}_3$ is incrementally Lagrangian-objective, and the tensor $\Delta t \mathbf{d}_3$ is incrementally Eulerian-objective.

If the tensors in the pairs $(\mathbf{d}, \mathbf{V})$ and $(\mathbf{D}, \mathbf{U})$ are coaxial and there are no rotations of the principal axes of the tensors $\mathbf{V}$ and $\mathbf{U}$ in the time interval $[t, t + \Delta t]$, then it follows from (2.13) that in this time interval,

$$\mathbf{D} = \overline{\ln \mathbf{U}}, \quad \mathbf{d} = \overline{\ln \mathbf{V}}.$$

In this case, the following equalities hold:

$$\Delta t \bar{\mathbf{d}}_3 = \int\limits_t^{t+\Delta t} \mathbf{D}\, d\tau, \quad \Delta t \mathbf{d}_3 = \int\limits_t^{t+\Delta t} \mathbf{d}\, d\tau,$$

i.e., for this type of deformations, the approximate incremental strains are equal to the exact values of incremental strains. Therefore, the incremental objective approximate tensors $\ln \mathbf{U}_r$ and $\ln \mathbf{V}_r$ can be considered the best approximations for incrementally Lagrangian- and Eulerian-objective incremental strains in the time interval $[t, t + \Delta t]$.

Remark 4.8 In fact, approximate incremental strains of the form $\Delta t \bar{\mathbf{d}}_3$ are proposed in [7, 9, 40]; however, in these papers it is recommended that for applications the approximations of the tensor $\ln \mathbf{U}_r$ given below should be used, rather than the tensor itself.

Due to the high computational cost of determining the tensors $\ln \mathbf{U}_r$ and $\ln \mathbf{V}_r$, in applications one can use more cost-effective expressions for the tensors $\Delta t \bar{\mathbf{d}}^a \equiv \Delta t^0 \mathbf{d}$ and $\Delta t \mathbf{d}^a \equiv \Delta t^1 \mathbf{d}$, which, in turn, approximate the tensors $\ln \mathbf{U}_r$ and $\ln \mathbf{V}_r$ with varying degrees of accuracy. In particular, the tensors $\Delta t \bar{\mathbf{d}}_1$ and $\Delta t \bar{\mathbf{d}}_2$ ($\Delta t \mathbf{d}_1$ and $\Delta t \mathbf{d}_2$) approximate the tensor $\Delta t \bar{\mathbf{d}}_3$ ($\Delta t \mathbf{d}_3$) up to second-order terms. We propose to approximate the Hencky incremental strain tensors $\ln \mathbf{U}_r$ and $\ln \mathbf{V}_r$ by the Mooney incremental strain tensors up to third-order terms:

$$\Delta t \bar{\mathbf{d}}_4 \equiv \mathbf{E}_r^M = \frac{1}{2}(\Delta t \bar{\mathbf{d}}_1 + \Delta t \bar{\mathbf{d}}_2), \quad \Delta t \mathbf{d}_4 \equiv \mathbf{e}_r^M = \frac{1}{2}(\Delta t \mathbf{d}_1 + \Delta t \mathbf{d}_2).$$

Here the tensors $\mathbf{E}_r^M$ and $\mathbf{e}_r^M$ are the (incrementally Lagrangian- and Eulerian-objective) *right and left Mooney incremental strain tensors*, respectively

$$\mathbf{E}_r^M \equiv \frac{1}{4}(\mathbf{F}_r^T \cdot \mathbf{F}_r - \mathbf{F}_r^{-1} \cdot \mathbf{F}_r^{-T})\left[= \frac{1}{2}(\mathbf{E}_r^{(2)} + \mathbf{E}_r^{(-2)})\right],$$

$$\mathbf{e}_r^M \equiv \frac{1}{4}(\mathbf{F}_r \cdot \mathbf{F}_r^T - \mathbf{F}_r^{-T} \cdot \mathbf{F}_r^{-1})\left[= \frac{1}{2}(\mathbf{e}_r^{(2)} + \mathbf{e}_r^{(-2)})\right].$$

As noted in Remark 4.8, incrementally Lagrangian-objective approximate incremental strains $\Delta t \bar{\mathbf{d}}^a$ that differ from the similar tensors considered above are proposed in [7, 9, 39]. The approximate incremental strain tensor

$$\Delta t \bar{\mathbf{d}}_N \equiv \left(\mathbf{I} - \frac{1}{2}\mathbf{E}_r^{(2)}\right) \cdot \mathbf{E}_r^{(2)} \cdot \left(\mathbf{I} - \frac{1}{2}\mathbf{E}_r^{(2)}\right)$$

is presented in [39], the tensor

$$\Delta t \bar{\mathbf{d}}_W \equiv 2(\mathbf{U}_r - \mathbf{I}) \cdot (\mathbf{U}_r + \mathbf{I})^{-1}$$

is used in [9], and the tensor

$$\Delta t \bar{\mathbf{d}}_R \equiv \mathbf{E}_r^{(-2)} + (\mathbf{E}_r^{(-2)})^2$$

is proposed in [7].

It can be shown that (like the tensor $\Delta t \bar{\mathbf{d}}_4$), the tensors $\Delta t \bar{\mathbf{d}}_N$, $\Delta t \bar{\mathbf{d}}_W$, and $\Delta t \bar{\mathbf{d}}_R$ approximate the tensor $\Delta t \bar{\mathbf{d}}_3$ ($= \ln \mathbf{U}_r$) up to third-order terms. However, compared to the first three tensors, the tensor $\mathbf{E}_r^M$ ($= \Delta t \bar{\mathbf{d}}_4$) more accurately reflects the behavior of the tensor $\ln \mathbf{U}_r$ at large time steps, since unlike the tensors $\Delta t \bar{\mathbf{d}}_N$, $\Delta t \bar{\mathbf{d}}_W$, and $\Delta t \bar{\mathbf{d}}_R$, both of the tensors $\ln \mathbf{U}_r$ and $\mathbf{E}_r^M$ belong to the family of symmetrically physical strain tensors (cf., [27]). In addition, the tensor $\Delta t \bar{\mathbf{d}}_W$ additionally requires the determination of the tensor $\mathbf{U}_r$ from the polar decomposition of the tensor $\mathbf{F}_r$.

All approximate incremental strains (and their objective counterparts) introduced above are listed in Table 4.1.

Table 4.1 Expressions for incrementally objective approximate incremental strains $\Delta t \bar{\mathbf{d}}^a$ and $\Delta t \mathbf{d}^a$ presented in the literature and in this paper

Tensor identification	Incremental objectivity type	Expression for strain	Literature reference[b]	Order of accuracy of approximation
$\Delta t \bar{\mathbf{d}}_1$	Lagrangian	$\mathbf{E}_r^{(2)} \equiv (\mathbf{F}_r^T \cdot \mathbf{F}_r - \mathbf{I})/2$	[4]	Second
$\Delta t \mathbf{d}_1$	Eulerian	$\mathbf{e}_r^{(2)} \equiv (\mathbf{F}_r \cdot \mathbf{F}_r^T - \mathbf{I})/2$	[32]	Second
$\Delta t \bar{\mathbf{d}}_2$	Lagrangian	$\mathbf{E}_r^{(-2)} \equiv$ $(\mathbf{I} - \mathbf{F}_r^{-1} \cdot \mathbf{F}_r^{-T})/2$	–	Second
$\Delta t \mathbf{d}_2$	Eulerian	$\mathbf{e}_r^{(-2)} \equiv$ $(\mathbf{I} - \mathbf{F}_r^{-T} \cdot \mathbf{F}_r^{-1})/2$	[4, 26, 33–35]	Second
$\Delta t \bar{\mathbf{d}}_3$	Lagrangian	$\ln \mathbf{U}_r$	[7, 9, 40]	Exact
$\Delta t \mathbf{d}_3$	Eulerian	$\ln \mathbf{V}_r$	–	Exact
$\Delta t \bar{\mathbf{d}}_4$	Lagrangian	$\mathbf{E}_r^M \equiv (\mathbf{F}_r^T \cdot \mathbf{F}_r - \mathbf{F}_r^{-1} \cdot \mathbf{F}_r^{-T})/4$	–	Third
$\Delta t \mathbf{d}_4$	Eulerian	$\mathbf{e}_r^M \equiv (\mathbf{F}_r \cdot \mathbf{F}_r^T - \mathbf{F}_r^{-T} \cdot \mathbf{F}_r^{-1})/4$	–	Third
$\Delta t \bar{\mathbf{d}}_N$	Lagrangian	$(\mathbf{I} - \frac{1}{2}\mathbf{E}_r^{(2)}) \cdot \mathbf{E}_r^{(2)} \cdot (\mathbf{I} - \frac{1}{2}\mathbf{E}_r^{(2)})$	[39]	Third
$\Delta t \mathbf{d}_N$	Eulerian	$(\mathbf{I} - \frac{1}{2}\mathbf{e}_r^{(2)}) \cdot \mathbf{e}_r^{(2)} \cdot (\mathbf{I} - \frac{1}{2}\mathbf{e}_r^{(2)})$	–	Third
$\Delta t \bar{\mathbf{d}}_W$	Lagrangian	$2(\mathbf{U}_r - \mathbf{I}) \cdot (\mathbf{U}_r + \mathbf{I})^{-1}$	[9]	Third
$\Delta t \mathbf{d}_W$	Eulerian	$2(\mathbf{V}_r - \mathbf{I}) \cdot (\mathbf{V}_r + \mathbf{I})^{-1}$	–	Third
$\Delta t \bar{\mathbf{d}}_R$	Lagrangian	$\mathbf{E}_r^{(-2)} + (\mathbf{E}_r^{(-2)})^2$	[7]	Third
$\Delta t \mathbf{d}_R$	Eulerian	$\mathbf{e}_r^{(-2)} + (\mathbf{e}_r^{(-2)})^2$	–	Third

[b]A dash (—) indicates that this expression is not presented in literature

4.3.3 Expressions for Approximate Incremental Rotations for Strong Incrementally Objective Algorithms

To obtain the approximate incremental rotation tensor $^{\Delta t}\Psi_\omega$, in the time interval $[t, t + \Delta t]$ it is necessary to solve Eq. (2.16), where the tensor ω is determined from the incremental counterpart of Eq. (2.46)$_2$. According to Proposition 2.2, we can first find the incremental rotation tensor $^{\Delta t}\Psi_v{}^1$ associated with the incremental counterpart of the vorticity tensor $\mathbf{w}$ and then determine the incremental rotation tensor associated with the incremental counterpart of the second skew-symmetric tensor on the r.h.s. of (2.46)$_2$. To determine the incremental rotation tensor $^{\Delta t}\Psi_v$, we use the expression on the r.h.s. of (2.12) assuming that in determining the polar spin tensor ω^R, the reference configuration coincides with the configuration of the deformable body $\mathfrak{B}$ at time t.

Using the generalized midpoint rule, we express the incremental counterpart of equality (2.12) ($\alpha \in [0, 1]$) as

$$\Delta t \, {}^\alpha\mathbf{w} = \Delta t \, {}^\alpha\omega^R + \sum_{i \neq j=1}^{{}^\alpha m} \frac{{}^\alpha\lambda_i - {}^\alpha\lambda_j}{{}^\alpha\lambda_i + {}^\alpha\lambda_j} {}^\alpha\mathbf{V}_i \cdot (\Delta t \, {}^\alpha\mathbf{d}) \cdot {}^\alpha\mathbf{V}_j. \tag{4.50}$$

Here the tensors $\Delta t \, {}^\alpha\mathbf{d}$ and $\Delta t \, {}^\alpha\mathbf{w}$ are defined in (4.15), the tensor $\Delta t \, {}^\alpha\omega^R$ is the incremental counterpart of the polar spin tensor ω^R defined in (2.8)$_4$; ${}^\alpha m, {}^\alpha\lambda_i, {}^\alpha\lambda_j$, ${}^\alpha\mathbf{V}_i$, and ${}^\alpha\mathbf{V}_j$ ($i, j = 1, \ldots, {}^\alpha m$) are the eigenindex, eigenvalues and subordinate eigenprojections of the tensor ${}^\alpha\mathbf{V}_r$ obtained from the polar decomposition of the tensor ${}^\alpha\mathbf{F}_r$. Since the configuration of the body $\mathfrak{B}$ at time t (i.e., at $\alpha = 0$) is assumed to be the reference configuration, it follows that ${}^0m = 1$ and the second term on the r.h.s. of (4.50) is equal to the zero tensor. At the same time, expression (2.13)$_2$ implies that all incrementally Eulerian-objective approximate incremental strain tensors $\Delta t \, {}^1\mathbf{d}(= \Delta t \mathbf{d}^a)$ given in Table 4.1 approximate only the component of the incremental strain tensor that is coaxial with the tensor $\mathbf{V}_r$ and ignore the component of the incremental strain tensor that is orthogonal to the tensor $\mathbf{V}_r$ (i.e., they ignore the counterpart of the second term on the r.h.s. of (2.13)$_2$ for this tensor). This implies that for any tensor $\Delta t \mathbf{d}^a$, the second term on the r.h.s. of (4.50) is equal to the zero tensor at time $t + \Delta t$ (i.e., at $\alpha = 1$). Ignoring the second term on the r.h.s. of (4.50) for the remaining values of $\alpha \in (0, 1)$, we obtain the following expression for the approximate incremental vorticity tensor:

$$\Delta t \, {}^\alpha\mathbf{w} = \Delta t \, {}^\alpha\omega^R \quad (\alpha \in [0, 1]). \tag{4.51}$$

Since, in the time interval $[t, t + \Delta t]$, the exact solution of the problem

$$\dot{\mathbf{R}} = \omega^R \cdot \mathbf{R}, \quad \mathbf{R} = \mathbf{I} \text{ at } t = t \tag{4.52}$$

---

[1] Since the images of the w and ω indices are similar, we use the notation Ψ_v instead of the notation Ψ_w to avoid confusion.

is the tensor $\mathbf{R}_r$ obtained from the polar decomposition of the tensor $\mathbf{F}_r$, it follows from (4.51) and (4.52) that within the framework of our approximations, the following expression for the incremental rotation tensor holds:

$$^{\Delta t}\mathbf{\Psi}_v = \mathbf{R}_r. \tag{4.53}$$

Remark 4.9 In [7, 9, 39, 40], the same expression for the tensor $^{\Delta t}\mathbf{\Psi}_v$ is used to integrate hypoelastic CRs based on the Zaremba–Jaumann stress rate, but no arguments are given for the choice of the incremental rotation tensor in the form (4.53). In contrast, we show that expression (4.53) for this tensor is compatible with the employed approximations of the incremental strain tensor.

According to Proposition (2.2) and expression $(2.46)_2$ for the tensor $\boldsymbol{\omega} \in \mathcal{T}^2_{\text{skew}}$, the tensor $\mathbf{\Psi}_\omega \in \mathcal{T}^{2+}_{\text{orth}}$ can be represented as the inner product of two tensors $\mathbf{\Psi}_v, \mathbf{\Psi}_2 \in \mathcal{T}^{2+}_{\text{orth}}$:

$$\mathbf{\Psi}_\omega = \mathbf{\Psi}_v \cdot \mathbf{\Psi}_2. \tag{4.54}$$

Here the tensor $\mathbf{\Psi}_v$ is found by solving the Cauchy problem (2.33) ($\boldsymbol{\omega}_1 = \mathbf{w}$), and the tensor $\mathbf{\Psi}_2$ by solving the Cauchy problem (2.35) taking into account expression (2.34) ($\mathbf{\Psi}_1 = \mathbf{\Psi}_v$) in which the tensor $\boldsymbol{\omega}_2 \in \mathcal{T}^2_{\text{skew}}$ is defined as

$$\boldsymbol{\omega}_2 = \sum_{i \neq j=1}^{m} g_{ij} \mathbf{V}_i \cdot \mathbf{d} \cdot \mathbf{V}_j. \tag{4.55}$$

Here m, λ_i, λ_j, $\mathbf{V}_i$, and $\mathbf{V}_j$ $(i, j = 1, \ldots, m)$ are the eigenindex, eigenvalues, and subordinate eigenprojections of the tensor $\mathbf{V}$ obtained from the polar decomposition of the tensor $\mathbf{F}$; $g_{ij} \equiv g(\lambda_i, \lambda_j)$ $(g_{ji} = -g_{ij})$.

Our next goal is to develop an algorithm for determining the incremental rotation tensor

$$^{\Delta t}\mathbf{\Psi}_\omega \equiv {}^{t+\Delta t}\mathbf{\Psi}_\omega \cdot {}^{t}\mathbf{\Psi}_\omega^T. \tag{4.56}$$

From (4.54) and (4.56), we obtain the following expression for the incremental rotation tensor:

$$^{\Delta t}\mathbf{\Psi}_\omega = {}^{t+\Delta t}\mathbf{\Psi}_v \cdot {}^{\Delta t}\mathbf{\Psi}_2 \cdot {}^{t}\mathbf{\Psi}_v^T, \quad ^{\Delta t}\mathbf{\Psi}_2 \equiv {}^{t+\Delta t}\mathbf{\Psi}_2 \cdot {}^{t}\mathbf{\Psi}_2^T. \tag{4.57}$$

Using the Rodrigues formula, we obtain the approximate tensor $^{\Delta t}_\alpha\mathbf{\Psi}_2$ associated with the skew-symmetric tensor $\Delta t \, {}^\alpha\tilde{\boldsymbol{\omega}}_2$—the incremental counterpart of the tensor $\tilde{\boldsymbol{\omega}}_2$ defined in (2.34)

$$\Delta t \, {}^\alpha\tilde{\boldsymbol{\omega}}_2 \equiv {}^{t+\alpha\Delta t}\mathbf{\Psi}_v^T \cdot (\Delta t \, {}^\alpha\boldsymbol{\omega}_2) \cdot {}^{t+\alpha\Delta t}\mathbf{\Psi}_v.$$

Using the exponential representation of the tensor $^{\Delta t}_\alpha\mathbf{\Psi}_2$ (see, e.g., [6]) and Proposition (2.4), we arrive at the equality

$$\overset{\Delta t}{\alpha}\boldsymbol{\Psi}_2 = {}^{t+\alpha\Delta t}\boldsymbol{\Psi}_v^T \cdot \overset{\Delta t}{\alpha}\boldsymbol{\Psi}_{\omega_2} \cdot {}^{t+\alpha\Delta t}\boldsymbol{\Psi}_v, \tag{4.58}$$

where the tensor $\overset{\Delta t}{\alpha}\boldsymbol{\Psi}_{\omega_2} \in \mathcal{T}_{\text{orth}}^{2+}$ is obtained using the Rodrigues formula and is associated with the skew-symmetric tensor $\Delta t\,{}^\alpha\boldsymbol{\omega}_2$ — the incremental counterpart of the tensor $\boldsymbol{\omega}_2$ defined by (4.55)

$$\Delta t^\alpha\boldsymbol{\omega}_2 = \sum_{i\neq j=1}^{{}^{t+\alpha\Delta t}m} {}^{t+\alpha\Delta t}g_{ij} {}^{t+\alpha\Delta t}\mathbf{V}_i \cdot (\Delta t^\alpha\mathbf{d}) \cdot {}^{t+\alpha\Delta t}\mathbf{V}_j.$$

Here ${}^{t+\alpha\Delta t}m$, ${}^{t+\alpha\Delta t}\lambda_i$, ${}^{t+\alpha\Delta t}\lambda_j$, ${}^{t+\alpha\Delta t}\mathbf{V}_i$, ${}^{t+\alpha\Delta t}\mathbf{V}_j$ $(i, j = 1, \ldots, {}^{t+\alpha\Delta t}m)$ are the eigenindex, eigenvalues, and subordinate eigenprojections of the left stretch tensor ${}^{t+\alpha\Delta t}\mathbf{V}$ obtained from the polar decomposition of the deformation gradient tensor ${}^{t+\alpha\Delta t}\mathbf{F}$; ${}^{t+\alpha\Delta t}g_{ij} \equiv g({}^{t+\alpha\Delta t}\lambda_i, {}^{t+\alpha\Delta t}\lambda_j)$; $\Delta t^\alpha\mathbf{d}$ is the approximate incremental strain tensor.

Below we use two values of the parameter α ($\alpha = 0, 1$) which are compatible with the strong incrementally objective algorithms for integrating CRs for hypoelastic models. For brevity, as in Sect. 4.3.2, the quantities associated with time t ($\alpha = 0$) will be denoted by symbols with bars above them, and the quantities associated with time $t + \Delta t$ ($\alpha = 1$) will be denoted by symbols without bars.

We first assign a value $\alpha = 1$ to the parameter α. Using equalities (4.57) and (4.58) and taking into account the equality

$$\overset{\Delta t}{}\boldsymbol{\Psi}_v \equiv {}^{t+\Delta t}\boldsymbol{\Psi}_v \cdot {}^t\boldsymbol{\Psi}_v^T = \mathbf{R}_r, \tag{4.59}$$

we obtain the following expression for the tensor $\overset{\Delta t}{}\boldsymbol{\Psi}_\omega$:

$$\overset{\Delta t}{}\boldsymbol{\Psi}_\omega = \overset{\Delta t}{}\boldsymbol{\Psi}_{\omega_2} \cdot \mathbf{R}_r \quad \text{for} \quad \alpha = 1. \tag{4.60}$$

Here the incremental rotation tensor $\overset{\Delta t}{}\boldsymbol{\Psi}_{\omega_2}$ is determined using the Rodrigues formula and is associated with the spin tensor

$$\Delta t\boldsymbol{\omega}_2 = \sum_{i\neq j=1}^{{}^{t+\Delta t}m} {}^{t+\Delta t}g_{ij} {}^{t+\Delta t}\mathbf{V}_i \cdot (\Delta t\mathbf{d}^a) \cdot {}^{t+\Delta t}\mathbf{V}_j, \tag{4.61}$$

where $\Delta t\mathbf{d}^a$ is any of the expressions for the incrementally Eulerian-objective approximate incremental strain tensor $\Delta t\mathbf{d}^a$ given in Table 4.1.

We now assign a value $\alpha = 0$ to the parameter α. Using equalities (4.57)–(4.59), we obtain

$$\overset{\Delta t}{}\bar{\boldsymbol{\Psi}}_\omega = \mathbf{R}_r \cdot \overset{\Delta t}{}\bar{\boldsymbol{\Psi}}_{\bar{\omega}_2} \quad \text{for} \quad \alpha = 0. \tag{4.62}$$

Here the incremental rotation tensor $\overset{\Delta t}{}\bar{\boldsymbol{\Psi}}_{\bar{\omega}_2}$ is determined using the Rodrigues formula and is associated with the spin tensor

$$\Delta t \bar{\omega}_2 = \sum_{i \neq j=1}^{{}^t m} {}^t g_{ij} {}^t \mathbf{V}_i \cdot (\Delta t \bar{\mathbf{d}}^a) \cdot {}^t \mathbf{V}_j, \tag{4.63}$$

where $\Delta t \bar{\mathbf{d}}^a$ is any of the expressions for the incrementally Lagrangian-objective incremental approximate strain tensor given in Table 4.1.

In particular, when using CRs for hypoelastic models based on the Zaremba–Jaumann stress rate, we have $\Delta t \omega_2 = \Delta t \bar{\omega}_2 = \mathbf{0}$, whence ${}^{\Delta t} \boldsymbol{\Psi}_{\omega_2} = {}^{\Delta t} \bar{\boldsymbol{\Psi}}_{\bar{\omega}_2} = \mathbf{I}$. Then it follows from (4.60) and (4.62) that for such hypoelastic models, the following equalities hold:

$$^{\Delta t} \boldsymbol{\Psi}_\omega = {}^{\Delta t} \bar{\boldsymbol{\Psi}}_\omega (= {}^{\Delta t} \boldsymbol{\Psi}_v) = \mathbf{R}_r. \tag{4.64}$$

We now check whether the tensor ${}^{\Delta t} \boldsymbol{\Psi}_\omega$ has incremental objectivity of any type. Since the tensors ${}^{t+\Delta t} \mathbf{V}_i$ and ${}^{t+\Delta t} \mathbf{V}_j$ $(i, j = 1, \ldots, {}^{t+\Delta t} m)$ are Eulerian tensors associated with time $t + \Delta t$, they are simultaneously incremental Eulerian-objective tensors (see Chap. 3); the tensors $\Delta t \mathbf{d}^a$ are incremental Eulerian-objective tensors (see Sect. 4.3.2). Then expression on the r.h.s. of (4.61) implies that the tensor $\Delta t \omega_2$ is an incremental Eulerian-objective tensor. Since the proper orthogonal tensor ${}^{\Delta t} \boldsymbol{\Psi}_{\omega_2}$ is determined from the Rodrigues formula and is associated with the skew-symmetric tensor $\Delta t \omega_2$, it can be shown, using Proposition 2.4, that the tensor ${}^{\Delta t} \boldsymbol{\Psi}_{\omega_2}$ is also incremental Eulerian objective. Since the tensor $\mathbf{R}_r$ is incremental Eulerian–Lagrangian-objective, it follows from (4.60) that the tensor ${}^{\Delta t} \boldsymbol{\Psi}_\omega$ is incremental Eulerian–Lagrangian-objective.

We now check the existence of any incremental objectivity of the tensor ${}^{\Delta t} \bar{\boldsymbol{\Psi}}_\omega$. Since the tensors ${}^t \mathbf{V}_i$ and ${}^t \mathbf{V}_j$ $(i, j = 1, \ldots, {}^t m)$ are Eulerian tensors associated with time t, they are simultaneously incremental Lagrangian-objective tensors (see Chap. 3); the tensors $\Delta t \bar{\mathbf{d}}^a$ are incremental Lagrangian-objective tensors (see Sect. 4.3.2). Then, using expression on the r.h.s. of (4.63), it can be shown that the tensor $\Delta t \bar{\omega}_2$ is an incremental Lagrangian-objective tensor. Since the proper orthogonal tensor ${}^{\Delta t} \bar{\boldsymbol{\Psi}}_{\bar{\omega}_2}$ associated with the skew-symmetric tensor $\Delta t \bar{\omega}_2$ is determined from the Rodrigues formula, it can be shown, using Proposition 2.4, that the tensor ${}^{\Delta t} \bar{\boldsymbol{\Psi}}_{\bar{\omega}_2}$ is also incremental Lagrangian-objective. Since the tensor $\mathbf{R}_r$ is incremental Eulerian–Lagrangian-objective, it follows from (4.62) that the tensor ${}^{\Delta t} \bar{\boldsymbol{\Psi}}_\omega$ is incremental Eulerian–Lagrangian-objective.

4.3.4 Expressions for Determining the Kirchhoff Stress Tensors

The Kirchhoff stress tensor $\boldsymbol{\tau}$ is found by step-by-step integration of the Cauchy problem (2.43). Using the representation for $\boldsymbol{\tau}^\omega$ in the form (2.38)$_2$, we rewrite Eq. (2.43)$_1$ as

$$\overline{\boldsymbol{\Psi}^{ET} \cdot \dot{\boldsymbol{\tau}} \cdot \boldsymbol{\Psi}^E} = \boldsymbol{\Psi}^{ET} \cdot (2\mu \mathbf{d}) \cdot \boldsymbol{\Psi}^E. \tag{4.65}$$

Using the backward finite difference approximation (at time $t + \Delta t$) of the l.h.s. of (4.65), we obtain the following expression for the tensor $^{t+\Delta t}\boldsymbol{\tau}$:

$$^{t+\Delta t}\boldsymbol{\tau} = {}^{\Delta t}\boldsymbol{\Psi}_\omega \cdot {}^t\boldsymbol{\tau} \cdot {}^{\Delta t}\boldsymbol{\Psi}_\omega^T + 2\mu\Delta t\,\mathbf{d}^a. \tag{4.66}$$

Here the tensor $^{\Delta t}\boldsymbol{\Psi}_\omega$ is determined using the algorithm presented in Sect. 4.3.3, and for the tensor $\Delta t \mathbf{d}^a$, we can use any of the expressions presented in Table 4.1.

Using the forward finite difference approximation (at time t) of the l.h.s. of (4.65), we obtain the following expression for the tensor $^{t+\Delta t}\boldsymbol{\tau}$:

$$^{t+\Delta t}\boldsymbol{\tau} = {}^{\Delta t}\bar{\boldsymbol{\Psi}}_\omega \cdot {}^t\boldsymbol{\tau} \cdot {}^{\Delta t}\bar{\boldsymbol{\Psi}}_\omega^T + 2\mu\,{}^{\Delta t}\bar{\boldsymbol{\Psi}}_\omega \cdot (\Delta t \bar{\mathbf{d}}^a) \cdot {}^{\Delta t}\bar{\boldsymbol{\Psi}}_\omega^T. \tag{4.67}$$

Here the tensor $^{\Delta t}\bar{\boldsymbol{\Psi}}_\omega$, as for expression (4.66), is determined using the algorithm presented in Sect. 4.3.3, and for the tensor $\Delta t \bar{\mathbf{d}}^a$ we can use any of the expressions presented in Table 4.1.

Remark 4.10 We approximate the tensor $\mathbf{e}_{incr} \equiv \int_t^{t+\Delta t} \mathbf{d}\,d\tau$ in terms of the tensor $\Delta t \bar{\mathbf{d}}^a$ at time t and the tensor $\Delta t \mathbf{d}^a$ at time $t + \Delta t$, which, under the assumption of generalized midpoint approximations, are related by the equality

$$\Delta t \mathbf{d}^a = \mathbf{R}_r \cdot (\Delta t \bar{\mathbf{d}}^a) \cdot \mathbf{R}_r^T.$$

In deriving equalities (4.66) and (4.67) from equality (4.65), we assumed that

$$\Delta t\,^{t+\Delta t}\mathbf{d} \approx \Delta t \mathbf{d}^a, \quad \Delta t\,^t\mathbf{d} \approx \Delta t \bar{\mathbf{d}}^a. \tag{4.68}$$

Approximations (4.68) generally do not ensure that expressions (4.66) and (4.67) are equivalent. However, for hypoelastic material models based on the Zaremba–Jaumann stress rate, the equalities (see (4.64)) $^{\Delta t}\boldsymbol{\Psi}_\omega = {}^{\Delta t}\bar{\boldsymbol{\Psi}}_\omega\,(= {}^{\Delta t}\boldsymbol{\Psi}_v) = \mathbf{R}_r$ hold; therefore, for these material models, expressions (4.66) and (4.67) are equivalent. For other material models based on corotational stress rates associated with spin tensors dependent on the choice of the reference configuration, expressions (4.66) and (4.67) will lead to similar results only for fairly small time steps (see Figs. 6.14, 6.15 and 6.16).

The incremental Eulerian–Lagrangian objectivity of the tensors $^{\Delta t}\boldsymbol{\Psi}_\omega$ and $^{\Delta t}\bar{\boldsymbol{\Psi}}_\omega$, the incremental Eulerian objectivity of the tensor (tensors) $\Delta t \mathbf{d}^a$, the incremental Lagrangian objectivity of the tensor (tensors) $\Delta t \bar{\mathbf{d}}^a$, and the incremental Lagrangian objectivity of the tensor $^t\boldsymbol{\tau}$ ensure the incremental (and hence absolute) Eulerian objectivity of the tensor $^{t+\Delta t}\boldsymbol{\tau}$ when using both expressions—(4.66) and (4.67). This implies the strong incremental objectivity of the algorithms developed in this section to integrate CRs for hypoelastic models. In other words, these algorithms exactly reproduce SRBMs of deformable bodies. However, expression (4.66) requires fewer arithmetic operations compared to expression (4.67); therefore, to integrate hypoelastic CRs, it is preferable to use the strong incrementally objective algorithm based on expression (4.66) for updating the stress tensor, rather than the similar algorithm based on expression (4.67).

Note that flowcharts of strong incrementally Lagrangian- and Eulerian-objective algorithms are given in Figs. A.3 and A.4, respectively.

References

1. T.J.R. Hughes, J. Winget, Int. J. Numer. Methods Eng. **15**, 1862 (1980). https://doi.org/10.1002/nme.1620151210
2. S.H. Lo, Int. J. Numer. Methods Eng. **26**, 121 (1988). https://doi.org/10.1002/nme.1620260109
3. P.M. Pinsky, M. Ortiz, K.S. Pister, Comput. Methods Appl. Mech. Eng. **40**, 137 (1983). https://doi.org/10.1016/0045-7825(83)90087-7
4. J.C. Simo, T.J.R. Hughes, *Computational Inelasticity* (Springer, N.Y., 1998)
5. A. Rodriguez-Ferran, P. Pegon, A. Huerta, Int. J. Numer. Methods Eng. **40**, 4363 (1997). https://doi.org/10.1002/(SICI)1097-0207(19971215)40:23<4363::AID-NME263>3.0.CO;2-Z
6. R. Rubinstein, S.N. Atluri, Comput. Methods Appl. Mech. Eng. **36**, 277 (1983). https://doi.org/10.1016/0045-7825(83)90125-1
7. M.M. Rashid, Int. J. Numer. Methods Eng. **36**, 3937 (1993). https://doi.org/10.1002/nme.1620362302
8. M.B. Rubin, O. Papes, J. Mech. Mater. Struct. **6**, 529 (2011). https://doi.org/10.2140/jomms.2011.6.529
9. G.G. Weber, A.M. Lush, A. Zavaliangos, L. Anand, Int. J. Plast. **6**, 701 (1990). https://doi.org/10.1016/0749-6419(90)90040-L
10. M.A. Crisfield, *Non-linear Finite Element Analysis of Solids and Structures: Vol. 2. Advanced Topics* (Wiley, Chichester, 1997)
11. R. de Borst, M.A. Crisfield, J.J.C. Remmers, C.V. Verhoosel, *Non-linear Finite Element Analysis of Solids and Structures*, 2nd edn. (Wiley, Chichester, 2012)
12. S.W. Key, R.D. Krieg, Comput. Methods Appl. Mech. Eng. **33**, 439 (1982). https://doi.org/10.1016/0045-7825(82)90118-9
13. T.J.R. Hughes, in *Theoretical Foundation for Large-scale Computations for Nonlinear Material Behavior*, eds. by S. Nemat-Nasser et al. (Martinus Nijhoff Publishers, Dordrecht, 1984), pp. 29–63
14. J. Argyris, Comput. Methods Appl. Mech. Eng. **32**, 85 (1982). https://doi.org/10.1016/0045-7825(82)90069-X
15. J. Ghaboussi, D.A. Pecknold, X.S. Wu, *Nonlinear Computational Solid Mechanics* (CRC Press, Boca Raton, 2017)
16. E.A. de Souza Neto, D. Peric, D.J.R. Owen, *Computational Methods for Plasticity: Theory and Applications* (Wiley, Chichester, 2008)
17. B.E. Healy, R.H. Dodds Jr., Comput. Mech. **9**, 95 (1992). https://doi.org/10.1007/BF00370065
18. S. Roy, A.F. Fossum, R.J. Dexter, Int. J. Eng. Sci. **30**, 119 (1992). https://doi.org/10.1016/0020-7225(92)90045-I
19. D.P. Flanagan, L.M. Taylor, Comput. Methods Appl. Mech. Eng. **62**, 305 (1987). https://doi.org/10.1016/0045-7825(87)90065-X
20. H. Xiao, O.T. Bruhns, A. Meyers, J. Elast. **47**, 51 (1997). https://doi.org/10.1023/A:1007356925912
21. H. Xiao, O.T. Bruhns, A. Meyers, J. Elast. **56**, 59 (1999). https://doi.org/10.1023/A:1007677619913

22. H. Xiao, O.T. Bruhns, A. Meyers, Acta Mech. **138**, 31 (1999). https://doi.org/10.1007/BF01179540
23. S.N. Korobeynikov, Arch. Appl. Mech. **90**, 313 (2020). https://doi.org/10.1007/s00419-019-01611-3
24. S.N. Korobeynikov, J. Elast. **143**, 147 (2021). https://doi.org/10.1007/s10659-020-09808-2
25. S.N. Korobeynikov, J. Elast. **93**, 105 (2008). https://doi.org/10.1007/s10659-008-9166-0
26. X. Zhou, K.K. Tamma, Finite Elem. Anal. Des. **39**, 783 (2003). https://doi.org/10.1016/S0168-874X(03)00059-3
27. S.N. Korobeynikov, J. Elast. **136**, 159 (2019). https://doi.org/10.1007/s10659-018-9699-9
28. O.T. Bruhns, H. Xiao, A. Meyers, Acta Mech. **155**, 95 (2002). https://doi.org/10.1007/BF01170842
29. O.T. Bruhns, A. Meyers, H. Xiao, Proc. R. Soc. A **460**, 909 (2004). https://doi.org/10.1098/rspa.2003.1184
30. K.W. Reed, S.N. Atluri, Comput. Methods Appl. Mech. Eng. **39**, 245 (1983). https://doi.org/10.1016/0045-7825(83)90094-4
31. K.W. Reed, S.N. Atluri, Int. J. Plast. **1**, 63 (1985). https://doi.org/10.1016/0749-6419(85)90014-2
32. M. Hollenstein, M. Jabareen, M.B. Rubin, Comput. Mech. **52**, 649 (2013). https://doi.org/10.1007/s00466-013-0838-7
33. M. Jabareen, Int. J. Eng. Sci. **96**, 46 (2015). https://doi.org/10.1016/j.ijengsci.2015.07.001
34. M. Kroon, M.B. Rubin, Finite Elem. Anal. Des. **177**, 103422 (2020). https://doi.org/10.1016/j.finel.2020.103422
35. M.B. Rubin, Finite Elem. Anal. Des. **175**, 103409 (2020). https://doi.org/10.1016/j.finel.2020.103409
36. J. Dabounou, J. Eng. Mech. **142**, 04016056 (2016). https://doi.org/10.1061/(ASCE)EM.1943-7889.0001112
37. M.S. Gadala, J. Wang, Finite Elem. Anal. Des. **35**, 379 (2000). https://doi.org/10.1016/S0168-874X(00)00003-2
38. M. Kleiber, P. Kowalczyk, *Introduction to Nonlinear Thermomechanics of Solids* (Springer, Switzerland, 2016)
39. J.C. Nagtegaal, Comput. Methods Appl. Mech. Eng. **33**, 469 (1982). https://doi.org/10.1016/0045-7825(82)90120-7
40. J.C. Nagtegaal, F.E. Veldpaus, in *Numerical Methods in Industrial Forming Processes*, ed. by J. Pittman (Wiley, Swansea, 1984), pp. 351–371
41. A. Rodriguez-Ferran, A. Huerta, J. Eng. Mech. **124**, 939 (1998). https://doi.org/10.1061/(ASCE)0733-9399(1998)124:9(939)

Chapter 5
Absolutely Objective Algorithms for Integrating CRs for Hooke-Like Hypoelastic Models

Abstract In this section, unlike in Chap. 4, where the incremental tensors $\mathbf{R}_r$, $\mathbf{U}_r$, and $\mathbf{V}_r$ were used, we develop absolutely objective algorithms to integrate CRs for hypoelastic models using the incremental tensors $^{\Delta t}\mathbf{R}$, $^{\Delta t}\mathbf{U}$, and $^{\Delta t}\tilde{\mathbf{V}}$ defined in Chap. 3 in the time interval $[t, t + \Delta t]$. Absolutely Lagrangian-objective algorithms are presented in Sect. 5.1, and absolutely Eulerian-objective for integrating CRs for hypoelastic models in Sect. 5.2. Furthermore, a number of absolutely objective approximate incremental strain tensors are generated, and their advantages and disadvantages are analyzed in Chap. 6 by solving the simple extension and simple shear problems. In Sect. 5.3, a comparative analysis of both theoretically equivalent (Lagrangian and Eulerian) versions in application to the numerical integration of hypoelastic CRs is performed, showing that the Lagrangian version is preferred.

5.1 Absolutely Lagrangian-Objective Algorithm

As all other algorithms discussed in this book, the absolutely Lagrangian-objective algorithm includes three main steps: (1) determining the absolutely Lagrangian-objective approximate incremental strain tensor (Sect. 5.1.1), (2) finding the absolutely Lagrangian-objective approximate incremental rotation tensor (Sect. 5.1.2), and (3) determining the absolutely Lagrangian-objective approximate incremental rotated Kirchhoff stress tensor (Sect. 5.1.3).

5.1.1 Absolutely Lagrangian-Objective Approximate Incremental Strain Tensors

Definition 5.1 We define the *absolutely Lagrangian-objective incremental strain tensor* as a tensor of the form

© The Author(s), under exclusive license to Springer Nature Switzerland AG 2023   53
S. Korobeynikov and A. Larichkin, *Objective Algorithms for Integrating Hypoelastic Constitutive Relations Based on Corotational Stress Rates*, SpringerBriefs in Continuum Mechanics, https://doi.org/10.1007/978-3-031-29632-1_5

$$\mathbf{E}_{\mathrm{incr}} \equiv \int\limits_{t}^{t+\Delta t} \mathbf{D}\, d\tau.$$

Considering expression $(2.10)_2$ at time $t + \Delta t$ and using the first-order backward finite difference approximation of the rate of the right stretch tensor $\dot{\mathbf{U}}$, we obtain the following approximation of the tensor $\mathbf{E}_{\mathrm{incr}}$:

$$\mathbf{E}_{\mathrm{incr}} \approx \frac{1}{2}[(^{t+\Delta t}\mathbf{U} - {}^{t}\mathbf{U}) \cdot {}^{t+\Delta t}\mathbf{U}^{-1} + {}^{t+\Delta t}\mathbf{U}^{-1} \cdot (^{t+\Delta t}\mathbf{U} - {}^{t}\mathbf{U})]$$

$$= \mathbf{I} - \frac{1}{2}(^{t}\mathbf{U} \cdot {}^{t+\Delta t}\mathbf{U}^{-1} + {}^{t+\Delta t}\mathbf{U}^{-1} \cdot {}^{t}\mathbf{U}).$$

Using the definition of the tensor $^{\Delta t}\mathbf{U}$ in (3.7), we obtain the following expression for the *absolutely Lagrangian-objective approximate incremental strain tensor*:

$$\Delta t \mathbf{D}_1 \equiv \mathbf{I} - \frac{1}{2}(^{\Delta t}\mathbf{U}^{-1} + {}^{\Delta t}\mathbf{U}^{-T}). \tag{5.1}$$

We call the tensor defined on the r.h.s. of (5.1) the *absolutely Lagrangian-objective incremental Hill strain tensor* (see Table 2.1).

Remark 5.1 The method of introducing the tensor $\Delta t \mathbf{D}_1$ shows that this absolutely Lagrangian-objective approximate incremental strain tensor is associated with time $t + \Delta t$. The same rule is valid for all other tensors of this form that are introduced in this section.

To introduce the next absolutely Lagrangian-objective approximate incremental strain tensor, we use the equality (see, e.g., Eq. $(125)_1$ in [1])

$$\dot{\mathbf{E}}^{(2)} = \mathbf{U} \cdot \mathbf{D} \cdot \mathbf{U}. \tag{5.2}$$

Considering equality (5.2) at time $t + \Delta t$ and using the first-order backward finite difference approximation of the tensor $\dot{\mathbf{E}}^{(2)}$, we obtain the following approximation of the tensor $\mathbf{E}_{\mathrm{incr}}$:

$$\Delta t \mathbf{D}_2 \equiv {}^{t+\Delta t}\mathbf{U}^{-1} \cdot (^{t+\Delta t}\mathbf{E}^{(2)} - {}^{t}\mathbf{E}^{(2)}) \cdot {}^{t+\Delta t}\mathbf{U}^{-1}. \tag{5.3}$$

Using the definition of the tensor $\mathbf{E}^{(2)}$ (see Table 2.1), we obtain the equality

$$^{t+\Delta t}\mathbf{E}^{(2)} - {}^{t}\mathbf{E}^{(2)} = \frac{1}{2}(^{t+\Delta t}\mathbf{U} \cdot {}^{t+\Delta t}\mathbf{U} - {}^{t}\mathbf{U} \cdot {}^{t}\mathbf{U}). \tag{5.4}$$

Substituting expression (5.4) into the r.h.s. of (5.3) and using the definition of the tensor $^{\Delta t}\mathbf{U}$ in (3.7), we obtain the following expression for the absolutely Lagrangian-objective approximate incremental strain tensor:

$$\Delta t \mathbf{D}_2 = \frac{1}{2}(\mathbf{I} - {}^{\Delta t}\mathbf{U}^{-T} \cdot {}^{\Delta t}\mathbf{U}^{-1}). \qquad (5.5)$$

We call the tensor expressed by the r.h.s. of (5.5) the *absolutely Lagrangian-objective incremental Karni–Reiner strain tensor* (see Table 2.1).

The absolutely Lagrangian-objective incremental strain tensor can be also approximated using the following equality (see, e.g., Eq. $(125)_2$ in [1]):

$$\dot{\mathbf{E}}^{(-2)} = \mathbf{U}^{-1} \cdot \mathbf{D} \cdot \mathbf{U}^{-1}. \qquad (5.6)$$

Considering equality (5.6) at time $t + \Delta t$ and using the first-order backward finite difference approximation of the tensor $\dot{\mathbf{E}}^{(-2)}$, we obtain the following approximation of the tensor $\mathbf{E}_{incr}$:

$$\Delta t \mathbf{D}_3 \equiv {}^{t+\Delta t}\mathbf{U} \cdot ({}^{t+\Delta t}\mathbf{E}^{(-2)} - {}^t\mathbf{E}^{(-2)}) \cdot {}^{t+\Delta t}\mathbf{U}. \qquad (5.7)$$

Using the definition of the tensor $\mathbf{E}^{(-2)}$ (see Table 2.1), we obtain the equality

$$ {}^{t+\Delta t}\mathbf{E}^{(-2)} - {}^t\mathbf{E}^{(-2)} = \frac{1}{2}({}^t\mathbf{U}^{-1} \cdot {}^t\mathbf{U}^{-1} - {}^{t+\Delta t}\mathbf{U}^{-1} \cdot {}^{t+\Delta t}\mathbf{U}^{-1}). \qquad (5.8)$$

Substituting expression (5.8) into the r.h.s. of (5.7) and using the definition of the tensor ${}^{\Delta t}\mathbf{U}$ in (3.7), we obtain the following expression for the absolutely Lagrangian-objective approximate incremental strain tensor:

$$\Delta t \mathbf{D}_3 = \frac{1}{2}({}^{\Delta t}\mathbf{U} \cdot {}^{\Delta t}\mathbf{U}^T - \mathbf{I}). \qquad (5.9)$$

We call the tensor expressed by the r.h.s. of (5.9) the *absolutely Lagrangian-objective incremental Green–Lagrange strain tensor* (see Table 2.1).

One more approximation of the absolutely Lagrangian-objective incremental strain tensor can be obtained as the average of the approximations $\Delta t \mathbf{D}_2$ and $\Delta t \mathbf{D}_3$

$$\Delta t \mathbf{D}_4 \equiv \frac{1}{2}(\Delta t \mathbf{D}_2 + \Delta t \mathbf{D}_3) = \frac{1}{4}({}^{\Delta t}\mathbf{U} \cdot {}^{\Delta t}\mathbf{U}^T - {}^{\Delta t}\mathbf{U}^{-T} \cdot {}^{\Delta t}\mathbf{U}^{-1}). \qquad (5.10)$$

We call the tensor expressed by the r.h.s. of (5.10) the *absolutely Lagrangian-objective incremental right Mooney strain tensor* (see Table 2.1).

We determine the symmetric component ${}^{\Delta t}\mathbf{U}^s$ of the tensor ${}^{\Delta t}\mathbf{U}$

$$ {}^{\Delta t}\mathbf{U}^s \equiv \frac{1}{2}({}^{\Delta t}\mathbf{U} + {}^{\Delta t}\mathbf{U}^T). $$

One more approximation of the absolutely Lagrangian-objective incremental strain tensor can be defined as

$$\Delta t \mathbf{D}_5 \equiv \ln {}^{\Delta t}\mathbf{U}^s. \qquad (5.11)$$

We call the tensor expressed by the r.h.s. of (5.11) the *absolutely Lagrangian-objective incremental right Hencky strain tensor* (see Table 2.1).

5.1.2 Absolutely Lagrangian-Objective Approximate Incremental Rotation Tensor

Definition 5.2 We define the *absolutely Lagrangian-objective incremental counterpart of the spin tensor* $\boldsymbol{\Omega}$ as a tensor of the form (see $(2.44)_1$ and (2.45))

$$\Delta t \boldsymbol{\Omega}^a \equiv \sum_{i \neq j = 1}^{m} r_{ij} \mathbf{U}_i \cdot (\Delta t \mathbf{D}^a) \cdot \mathbf{U}_j, \tag{5.12}$$

where $\Delta t \mathbf{D}^a$ is the absolutely Lagrangian-objective approximate incremental strain tensor and $r_{ij} \equiv r(\lambda_i, \lambda_j)$ and λ_i, λ_j and $\mathbf{U}_i$, $\mathbf{U}_j$ $(i, j = 1, \ldots, m)$ are the eigenvalues and subordinate eigenprojections of the tensor $^{t+\Delta t}\mathbf{U}$ obtained from the polar decomposition of the tensor $^{t+\Delta t}\mathbf{F}$.

Definition 5.3 We define the *absolutely Lagrangian-objective approximate incremental rotation tensor* as the tensor $^{\Delta t}\boldsymbol{\Psi}_\Omega \in \mathcal{T}_{\text{orth}}^{2+}$ which is associated with the spin tensor $\Delta t \boldsymbol{\Omega}^a \in \mathcal{T}_{\text{skew}}^2$ and obtained by approximately solving problem (2.22) in the time interval $[t, t + \Delta t]$.

In this book, we use the Rodrigues formula (cf., [2, 3] and Appendix C) to approximately solve problem (2.22).

Remark 5.2 For the Lagrangian version of the G-model based on the Green-Naghdi stress rate, we have $r_{ij} = 0$ $(i, j = 1, \ldots, m)$. Therefore, for this material model, the expressions $\Delta t \boldsymbol{\Omega}^a = \mathbf{0}$ and $^{\Delta t}\boldsymbol{\Psi}_\Omega = \mathbf{I}$ hold.

5.1.3 Expression for Determining the Absolutely Lagrangian-Objective Rotated Kirchhoff Stress Tensor

The Cauchy problem (2.42) is solved numerically by step-by-step integration with first-order accuracy using the backward finite difference method and expression $(2.38)_1$. Assuming that the deformation kinematics is specified and the value of the tensor $^t\bar{\boldsymbol{\tau}}$ at time t is known, we obtain the following approximate expression for the tensor $^{t+\Delta t}\bar{\boldsymbol{\tau}}$ at time $t + \Delta t$:

$$^{t+\Delta t}\bar{\boldsymbol{\tau}} = {}^{\Delta t}\boldsymbol{\Psi}_\Omega \cdot {}^t\bar{\boldsymbol{\tau}} \cdot {}^{\Delta t}\boldsymbol{\Psi}_\Omega^T + 2\mu(\Delta t \mathbf{D}^a), \tag{5.13}$$

where $\Delta t \mathbf{D}^a$ is the absolutely Lagrangian-objective approximate incremental strain tensor (any of the tensors of this type defined in Sect. 5.1.1.

Note that flowchart of absolutely Lagrangian-objective algorithm is given in Fig. A.5.

5.2 Absolutely Eulerian-Objective Algorithm

The absolutely Eulerian-objective algorithm is a counterpart of the absolutely Lagrangian-objective algorithm and also includes three main steps: (1) determining the absolutely Eulerian- objective approximate incremental strain tensor (Sect. 5.2.1), (2) finding the absolutely Eulerian-Lagrangian-objective approximate incremental rotation tensor (Sect. 5.2.2), and (3) determining the absolutely Eulerian-objective approximate incremental Kirchhoff stress tensor (Sect. 5.2.3).

5.2.1 Absolutely Eulerian-Objective Approximate Incremental Strain Tensors

Definition 5.4 We define the *absolutely Eulerian-objective incremental strain tensor* as a tensor of the form

$$\mathbf{e}_{\mathrm{incr}} \equiv \int\limits_{t}^{t+\Delta t} \mathbf{d}\, d\tau.$$

Remark 5.3 In Chap. 4, we did not provide an explicit definition of the Eulerian-objective incremental strain tensor, but in fact it is given by the r.h.s. of $(4.15)_1$ (see the expression in parentheses). Nevertheless, in this section, we obtain approximations of the tensor $\mathbf{e}_{\mathrm{incr}}$ that are different from the similar approximations obtained in Chap. 4. In this section, *absolutely Eulerian-objective approximate incremental strain tensors* are denoted by a tilde above the symbol to avoid confusion in notations.

We associate the tensors $\Delta t \tilde{\mathbf{d}}^a$ approximating the tensor $\mathbf{e}_{\mathrm{incr}}$ with time $t + \Delta t$ and define them as the Eulerian counterparts of the tensors $\Delta t \mathbf{D}^a$, i.e.,

$$\Delta t \tilde{\mathbf{d}}_i \equiv {}^{t+\Delta t}\mathbf{R} \cdot \Delta t \mathbf{D}_i \cdot {}^{t+\Delta t}\mathbf{R}^T \quad (i = 1, 2, \ldots, 5). \tag{5.14}$$

We obtain explicit expressions for the tensors $\Delta t \tilde{\mathbf{d}}_i$ ($i = 1, 2, \ldots, 5$) using equalities (3.8) and (5.14). For $i = 1$, from (5.1) we obtain the following expression for the absolutely Eulerian-objective approximate incremental strain tensor:

$$\Delta t \tilde{\mathbf{d}}_1 = \mathbf{I} - \frac{1}{2}({}^{\Delta t}\tilde{\mathbf{V}}^{-1} + {}^{\Delta t}\tilde{\mathbf{V}}^{-T}). \tag{5.15}$$

We call the tensor expressed by the r.h.s. of (5.15) the *absolutely Eulerian-objective incremental Swainger strain tensor* (see Table 2.1).

Setting $i = 2$, from (5.5) we obtain

$$\Delta t \tilde{\mathbf{d}}_2 = \frac{1}{2}(\mathbf{I} - {}^{\Delta t}\tilde{\mathbf{V}}^{-T} \cdot {}^{\Delta t}\tilde{\mathbf{V}}^{-1}). \tag{5.16}$$

We call the tensor expressed by the r.h.s. of (5.16) the *absolutely Eulerian-objective incremental Almansi strain tensor* (see Table 2.1).

For $i = 3$, from (5.9) we obtain

$$\Delta t \tilde{\mathbf{d}}_3 = \frac{1}{2}({}^{\Delta t}\tilde{\mathbf{V}} \cdot {}^{\Delta t}\tilde{\mathbf{V}}^T - \mathbf{I}). \tag{5.17}$$

We call the tensor expressed by the r.h.s. of (5.17) the *absolutely Eulerian-objective incremental Finger strain tensor* (see Table 2.1).

For $i = 4$, from (5.10) we obtain

$$\Delta t \tilde{\mathbf{d}}_4 = \frac{1}{2}(\Delta t \tilde{\mathbf{d}}_2 + \Delta t \tilde{\mathbf{d}}_3) = \frac{1}{4}({}^{\Delta t}\tilde{\mathbf{V}} \cdot {}^{\Delta t}\tilde{\mathbf{V}}^T - {}^{\Delta t}\tilde{\mathbf{V}}^{-T} \cdot {}^{\Delta t}\tilde{\mathbf{V}}^{-1}). \tag{5.18}$$

We call the tensor expressed by the r.h.s. of (5.18) the *absolutely Eulerian-objective incremental left Mooney strain tensor* (see Table 2.1).

We determine the symmetric component ${}^{\Delta t}\tilde{\mathbf{V}}^s$ of the tensor ${}^{\Delta t}\tilde{\mathbf{V}}$:

$$\Delta t \tilde{\mathbf{V}}^s \equiv \frac{1}{2}({}^{\Delta t}\tilde{\mathbf{V}} + {}^{\Delta t}\tilde{\mathbf{V}}^T). \tag{5.19}$$

Setting $i = 5$, from (5.11) and (5.19) we obtain

$$\Delta t \tilde{\mathbf{d}}_5 = \ln {}^{\Delta t}\tilde{\mathbf{V}}^s. \tag{5.20}$$

We call the tensor expressed by the r.h.s. of (5.20) the *absolutely Eulerian-objective incremental left Hencky strain tensor* (see Table 2.1).

5.2.2 Absolutely Eulerian–Lagrangian-Objective Approximate Incremental Rotation Tensor

Let the tensor $\boldsymbol{\Psi}_\omega$ be the *absolutely Eulerian–Lagrangian-objective rotation tensor* which is associated with the spin tensor $\boldsymbol{\omega}$ and is determined by solving the Cauchy problem (2.16). We define the *absolutely Eulerian–Lagrangian-objective approximate incremental rotation tensor* ${}^{\Delta t}\boldsymbol{\Psi}_\omega$ as follows:

$$\Delta t \boldsymbol{\Psi}_\omega \equiv {}^{t+\Delta t}\boldsymbol{\Psi}_\omega \cdot {}^t \boldsymbol{\Psi}_\omega^T. \tag{5.21}$$

Using equality (2.37) ($\boldsymbol{\Psi}_\omega = \mathbf{R} \cdot \boldsymbol{\Psi}_\Omega$), from (5.21) we obtain the following expression for the tensor $^{\Delta t}\boldsymbol{\Psi}_\omega$:

$$^{\Delta t}\boldsymbol{\Psi}_\omega = {}^{t+\Delta t}\mathbf{R} \cdot {}^{\Delta t}\boldsymbol{\Psi}_\Omega \cdot {}^{t}\mathbf{R}^T, \tag{5.22}$$

where $^{\Delta t}\boldsymbol{\Psi}_\Omega \equiv {}^{t+\Delta t}\boldsymbol{\Psi}_\Omega \cdot {}^{t}\boldsymbol{\Psi}_\Omega^T$ is the absolutely Lagrangian-objective approximate incremental rotation tensor defined in Sect. 5.1.2. To determine this tensor using the Rodrigues formula, it is required to determine the absolutely Lagrangian-objective approximate incremental spin tensor $\Delta t \boldsymbol{\Omega}^a$ defined in (5.12). To obtain this tensor, we first determine the absolutely Eulerian-objective approximate incremental spin tensor $\Delta t \tilde{\boldsymbol{\omega}}^a$ from the expression

$$\Delta t \tilde{\boldsymbol{\omega}}^a \equiv \sum_{i \neq j=1}^{m} r_{ij} \mathbf{V}_i \cdot (\Delta t \tilde{\mathbf{d}}^a) \cdot \mathbf{V}_j, \tag{5.23}$$

where $\Delta t \tilde{\mathbf{d}}^a$ is the absolutely Eulerian-objective approximate incremental strain tensor (any of the tensors of this type defined in Sect. 5.2.1), $r_{ij} \equiv r(\lambda_i, \lambda_j)$, λ_i, λ_j, $\mathbf{V}_i$, and $\mathbf{V}_j$ ($i, j = 1, \ldots, m$) are the eigenvalues and subordinate eigenprojections of the tensor $^{t+\Delta t}\mathbf{V}$ obtained from the polar decomposition of the tensor $^{t+\Delta t}\mathbf{F}$. Then the tensor $\Delta t \boldsymbol{\Omega}^a$ is obtained from the expression

$$\Delta t \boldsymbol{\Omega}^a = {}^{t+\Delta t}\mathbf{R}^T \cdot (\Delta t \tilde{\boldsymbol{\omega}}^a) \cdot {}^{t+\Delta t}\mathbf{R}. \tag{5.24}$$

Using expression (5.24), we determine the tensor $^{\Delta t}\boldsymbol{\Psi}_\Omega$ from the Rodrigues formula. Using the exponential representation of this tensor, equality (5.24), and Proposition 2.4, we arrive at the following expression for the tensor $^{\Delta t}\boldsymbol{\Psi}_\Omega$:

$$^{\Delta t}\boldsymbol{\Psi}_\Omega = {}^{t+\Delta t}\mathbf{R}^T \cdot {}^{\Delta t}\boldsymbol{\Psi}_{\tilde{\omega}} \cdot {}^{t+\Delta t}\mathbf{R}, \tag{5.25}$$

where the tensor $^{\Delta t}\boldsymbol{\Psi}_{\tilde{\omega}}$ associated with the spin tensor $\Delta t \tilde{\boldsymbol{\omega}}^a$ defined in (5.23) is determined using the Rodrigues formula. Then from (5.22) and (5.25), we obtain the final expression for the tensor $^{\Delta t}\boldsymbol{\Psi}_\omega$

$$^{\Delta t}\boldsymbol{\Psi}_\omega = {}^{\Delta t}\boldsymbol{\Psi}_{\tilde{\omega}} \cdot {}^{\Delta t}\mathbf{R}, \tag{5.26}$$

where the tensor $^{\Delta t}\mathbf{R}$ is defined in $(3.4)_2$.

Remark 5.4 For the G-model based on the Green–Naghdi stress rate, the equalities $\Delta t \tilde{\boldsymbol{\omega}}^a = \mathbf{0}$ and $^{\Delta t}\boldsymbol{\Psi}_{\tilde{\omega}} = \mathbf{I}$ hold. Then from (5.26), we obtain $^{\Delta t}\boldsymbol{\Psi}_\omega = {}^{\Delta t}\mathbf{R}$.

5.2.3 Expression for Determining the Absolutely Eulerian-Objective Kirchhoff Stress Tensor

The Cauchy problem (2.43) is solved numerically by step-by-step integration with first-order accuracy using the backward finite difference method and expression $(2.38)_2$. Assuming that the deformation kinematics is specified and the value of the tensor $^t\boldsymbol{\tau}$ at time t is known, we obtain the approximate expression for the tensor $^{t+\Delta t}\boldsymbol{\tau}$ at time $t + \Delta t$

$$^{t+\Delta t}\boldsymbol{\tau} = {}^{\Delta t}\boldsymbol{\Psi}_\omega \cdot {}^t\boldsymbol{\tau} \cdot {}^{\Delta t}\boldsymbol{\Psi}_\omega^T + 2\mu(\Delta t\tilde{\mathbf{d}}^a), \qquad (5.27)$$

where $\Delta t\tilde{\mathbf{d}}^a$ is the absolutely Eulerian-objective approximate incremental strain tensor (any of the tensors of this type defined in Sect. 5.2.1).

Note that flowchart of absolutely Eulerian-objective algorithm is given in Fig. A.6.

5.3 Discussion of Absolutely Objective Algorithms

The absolutely objective algorithms for integrating CRs for hypoelastic material models are theoretically equivalent, which is confirmed by the results of our computer simulations. However, it can be seen from the above expressions that the absolutely Eulerian-objective algorithm requires more operations than the absolutely Lagrangian-objective algorithm. Therefore, the absolutely Lagrangian-objective algorithm is preferred for integrating CRs for hypoelastic material models.

The expressions obtained in Sects. 5.1.1 and 5.2.1 for the absolutely objective approximate incremental strain tensors $\Delta t\mathbf{D}^a$ and $\Delta t\tilde{\mathbf{d}}^a$ are listed in Table 5.1. Note that in the absence of rotations of material fibers of the deformable body ($\mathbf{R} = \mathbf{I}$), the expressions for the absolutely objective approximate incremental strain tensors and incrementally objective approximate incremental strain tensors coincide with each other and the following equalities hold:

$$\Delta t\tilde{\mathbf{d}}^a = \Delta t\mathbf{d}^a = \Delta t\mathbf{D}^a = \Delta t\tilde{\mathbf{d}}^a.$$

For this deformation,

$$\Delta t\mathbf{D}_5 = \Delta t\tilde{\mathbf{d}}_5 = \ln {}^{\Delta t}\mathbf{U} = \ln {}^{\Delta t}\mathbf{V} = \mathbf{E}_{\text{incr}} = \mathbf{e}_{\text{incr}},$$

whence it follows that the incremental logarithmic strain tensor gives the exact expression for the incremental strain tensor, as shown in the last two rows and the last column of Table 5.1.

Table 5.1 Expressions for the absolutely objective approximate incremental strain tensors $\Delta t \mathbf{D}^a$ and $\Delta t \tilde{\mathbf{d}}^a$

Tensor notation	Absolute objectivity type	Expression for strain	Order of accuracy of approximation
$\Delta t \mathbf{D}_1$	Lagrangian	$\mathbf{I} - \frac{1}{2}({}^{\Delta t}\mathbf{U}^{-1} + {}^{\Delta t}\mathbf{U}^{-T})$	Second
$\Delta t \tilde{\mathbf{d}}_1$	Eulerian	$\mathbf{I} - \frac{1}{2}({}^{\Delta t}\tilde{\mathbf{V}}^{-1} + {}^{\Delta t}\tilde{\mathbf{V}}^{-T})$	Second
$\Delta t \mathbf{D}_2$	Lagrangian	$(\mathbf{I} - {}^{\Delta t}\mathbf{U}^{-T} \cdot {}^{\Delta t}\mathbf{U}^{-1})/2$	Second
$\Delta t \tilde{\mathbf{d}}_2$	Eulerian	$(\mathbf{I} - {}^{\Delta t}\tilde{\mathbf{V}}^{-T} \cdot {}^{\Delta t}\tilde{\mathbf{V}}^{-1})/2$	Second
$\Delta t \mathbf{D}_3$	Lagrangian	$({}^{\Delta t}\mathbf{U} \cdot {}^{\Delta t}\mathbf{U}^{T} - \mathbf{I})/2$	Second
$\Delta t \tilde{\mathbf{d}}_3$	Eulerian	$({}^{\Delta t}\tilde{\mathbf{V}} \cdot {}^{\Delta t}\tilde{\mathbf{V}}^{T} - \mathbf{I})/2$	Second
$\Delta t \mathbf{D}_4$	Lagrangian	$({}^{\Delta t}\mathbf{U} \cdot {}^{\Delta t}\mathbf{U}^{T} - {}^{\Delta t}\mathbf{U}^{-T} \cdot {}^{\Delta t}\mathbf{U}^{-1})/4$	Third
$\Delta t \tilde{\mathbf{d}}_4$	Eulerian	$({}^{\Delta t}\tilde{\mathbf{V}} \cdot {}^{\Delta t}\tilde{\mathbf{V}}^{T} - {}^{\Delta t}\tilde{\mathbf{V}}^{-T} \cdot {}^{\Delta t}\tilde{\mathbf{V}}^{-1})/4$	Third
$\Delta t \mathbf{D}_5$	Lagrangian	$\ln {}^{\Delta t}\mathbf{U}^{s}$	Exact
$\Delta t \tilde{\mathbf{d}}_5$	Eulerian	$\ln {}^{\Delta t}\tilde{\mathbf{V}}^{s}$	Exact

References

1. S.N. Korobeynikov, Acta Mech. **229**, 1061 (2018). https://doi.org/10.1007/s00707-017-1972-7
2. R. Rubinstein, S.N. Atluri, Comput. Methods Appl. Mech. Eng. **36**, 277 (1983). https://doi.org/10.1016/0045-7825(83)90125-1
3. J. Argyris, Comput. Methods Appl. Mech. Eng. **32**, 85 (1982). https://doi.org/10.1016/0045-7825(82)90069-X

Chapter 6
Comparative Analysis and Verification of Objective Algorithms

Abstract In Chap. 6 we perform a comparative analysis of all objective algorithms discussed in this book for integrating considered CRs by solving problems of uniform deformation of hypoelastic bodies. Sect. 6.1 presents exact solutions of simple extension and simple shear problems and estimates of the accuracy of the approximate incremental strains discussed in Chaps. 4 and 5 as applied to these problems. Sect. 6.2 provides background on computer simulations. We give the results of simulations using weak incrementally objective algorithms (Sect. 6.3), strong incrementally objective algorithms (Sect. 6.4), and absolutely objective algorithms (Sect. 6.5) to integrate CRs for hypoelastic models. In Sect. 6.6, we discuss the simulation results and provide recommendations for the use of the algorithms considered.

6.1 Problems of Uniform Deformation of Hypoelastic Bodies: Exact Solutions and Estimates of the Accuracy of Approximate Incremental Strains

For convenience, we number the material models as follows[1]:

- #41: C-model of hypoelasticity based on the Zaremba–Jaumann stress rate;
- #42: G-model of hypoelasticity based on the Green–Naghdi stress rate;
- #43: G-model of hypoelasticity based on the logarithmic stress rate;
- #46: C-model of hypoelasticity based on the Hill stress rate.

We solve the *simple extension* and *simple shear* problems of a square sample (cubic sample under plane strain) (Fig. 6.1) with specified deformation kinematics and zero initial stresses for the reference configuration of the sample at time t_0.

For the *simple extension* problem (Fig. 6.1a), we adopt the law of motion

$$x_1(t) = X_1 + k_1 t X_1, \quad x_2(t) = X_2, \quad x_3(t) = X_3 \quad t \in [0, 1]. \qquad (6.1)$$

---

[1] These hypoelastic material model numbers correspond to the material model numbers in our homemade FE code used to solve the problems presented in this paper.

© The Author(s), under exclusive license to Springer Nature Switzerland AG 2023   63
S. Korobeynikov and A. Larichkin, *Objective Algorithms for Integrating Hypoelastic Constitutive Relations Based on Corotational Stress Rates*, SpringerBriefs in Continuum Mechanics, https://doi.org/10.1007/978-3-031-29632-1_6

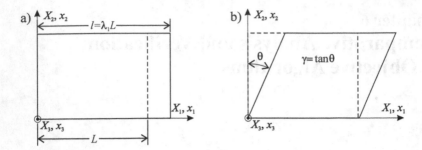

Fig. 6.1 Sketch of the kinematics in the simple extension (**a**) and simple shear (**b**) problems

Here X_i and x_i ($i = 1, 2, 3$) are the Cartesian coordinates of any material point of the sample in the reference (at time $t = 0$) and current (at time t) configurations, respectively, and the parameter k_1 is a constant coefficient. The principal stretches are obtained from the law of motion (6.1):

$$\lambda_1(t) = 1 + k_1 t, \quad \lambda_2(t) = \lambda_3(t) = 1.$$

Integrating CRs for hypoelastic models, we obtain the following values for the Cauchy stress tensor components in the material models considered ($\sigma_{12} = \sigma_{13} = \sigma_{23} = 0$):

- Material models ##41–43

$$\sigma_{11} = (\lambda + 2\mu) \ln \lambda_1 / \lambda_1, \tag{6.2}$$
$$\sigma_{22} = \sigma_{33} = \lambda \ln \lambda_1 / \lambda_1.$$

- Material model #46

$$\sigma_{11} = (\lambda + 2\mu)(\lambda_1 - 1)/\lambda_1, \tag{6.3}$$
$$\sigma_{22} = \sigma_{33} = \lambda(\lambda_1 - 1)/\lambda_1.$$

For the problem of *simple shear*, we adopt the following law of motion (see Fig. 6.1b):

$$x_1(t) = X_1 + \gamma(t)X_2, \quad x_2(t) = X_2, \quad x_3(t) = X_3 \quad t \in [0, 1]. \tag{6.4}$$

Here the *shear parameter* $\gamma(t)$ is a monotonically increasing function of time t of the form $\gamma(t) = k_2 t$, where the parameter k_2 is a constant coefficient.

Using the law of motion (6.4), we have the following expressions for the stretches [1]:

$$\lambda_1 = (\gamma + \sqrt{\gamma^2 + 4})/2, \quad \lambda_2 = (-\gamma + \sqrt{\gamma^2 + 4})/2 = \lambda_1^{-1}, \quad \lambda_3 = 1.$$

Note that in this solution, the deformation of the sample is isochoric ($J = \lambda_1\lambda_2\lambda_3 = 1$), and, therefore, the solutions for material models #41 and #46 coincide. Integrating the CRs for the material models considered, we obtain the following values for the Cauchy stress tensor components in the simple shear problem: ($\sigma_{13} = \sigma_{23} = \sigma_{33} = 0$):

- models #41 and #46 (see, e.g., [2–4])

$$\sigma_{11} = \mu(1 - \cos\gamma), \quad \sigma_{22} = -\sigma_{11}, \quad \sigma_{12} = \mu\sin\gamma. \tag{6.5}$$

- model #42 (see, e.g., [2–4])

$$\sigma_{11} = 4\mu[\cos(2\beta)\ln(\cos\beta) + \beta\sin(2\beta) - \sin^2\beta], \quad \sigma_{22} = -\sigma_{11}, \tag{6.6}$$
$$\sigma_{12} = 2\mu[2\beta\cos(2\beta) - 2\sin(2\beta)\ln(\cos\beta) - \cos(2\beta)\tan\beta].$$

Here we introduces the angle β, which is related to the shear parameter γ as $\gamma = 2\tan\beta$.
- model #43 (see, e.g., [1, 5–11])

$$\sigma_{11}(\gamma) = \frac{2\mu\gamma\ln\lambda_1(\gamma)}{\sqrt{4 + \gamma^2}}, \quad \sigma_{22} = -\sigma_{11}, \quad \sigma_{12}(\gamma) = \frac{4\mu\ln\lambda_1(\gamma)}{\sqrt{4 + \gamma^2}}. \tag{6.7}$$

Rewrite the motion laws (6.1) and (6.4) in vector-matrix form:

$$\mathbf{x}(t) = \mathbf{A}(t)\mathbf{X}, \quad \mathbf{x}(t) \equiv \begin{bmatrix} x_1(t) \\ x_2(t) \\ x_3(t) \end{bmatrix}, \quad \mathbf{X} \equiv \begin{bmatrix} X_1 \\ X_2 \\ X_3 \end{bmatrix}. \tag{6.8}$$

Here we introduced the matrix $\mathbf{A}(t)$ which has the form

$$\mathbf{A}(t) = \begin{bmatrix} 1 + k_1 t & 0 & 0 \\ 0 & 1 & 0 \\ 0 & 0 & 1 \end{bmatrix} \quad \text{or} \quad \mathbf{A}(t) = \begin{bmatrix} 1 & k_2 t & 0 \\ 0 & 1 & 0 \\ 0 & 0 & 1 \end{bmatrix}$$

for the laws of motion (6.1) and (6.4), respectively. Note that matrix $\mathbf{A}(t)$ coincides with the matrix representation of the tensor $\mathbf{F}$.

In addition to the laws of motion (6.8), we consider the same laws of motion with superimposed rotation in the form

$$\mathbf{x}^Q(t) \equiv \tilde{\mathbf{Q}}(t)\mathbf{x}(t) = \tilde{\mathbf{Q}}(t)\mathbf{A}(t)\mathbf{X}.$$

Hereinafter, $\tilde{\mathbf{Q}}(t)$ is the rotation matrix

$$\tilde{\mathbf{Q}}(t) \equiv \begin{bmatrix} \cos\theta(t) & \sin\theta(t) & 0 \\ -\sin\theta(t) & \cos\theta(t) & 0 \\ 0 & 0 & 1 \end{bmatrix},$$

where the rotation angle $\theta(t)$ is determined from the law of rotation with a constant rate of the rotation angle:

$$\theta(t) = k_3 2\pi t.$$

Here the parameter k_3 is a constant coefficient.

Let the stress tensor $\boldsymbol{\sigma}(t)$ be obtained by solving the simple extension or simple shear problems (i.e., the non-zero components of this tensor are given by expressions (6.2), (6.3) or (6.5)–(6.7)). Then, the stress tensor $\boldsymbol{\sigma}^Q(t)$ for deformations with superimposed rotation can be written in matrix form

$$\boldsymbol{\sigma}^Q(t) = \tilde{\mathbf{Q}}(t)\boldsymbol{\sigma}\tilde{\mathbf{Q}}^T(t), \quad \boldsymbol{\sigma} \equiv \begin{bmatrix} \sigma_{11} & \sigma_{12} & \sigma_{13} \\ \sigma_{12} & \sigma_{22} & \sigma_{23} \\ \sigma_{13} & \sigma_{23} & \sigma_{33} \end{bmatrix}. \tag{6.9}$$

The back rotated stress tensor $\bar{\boldsymbol{\sigma}}^Q(t)$ can be determined from the equality

$$\bar{\boldsymbol{\sigma}}^Q(t) = \tilde{\mathbf{Q}}^T(t)\boldsymbol{\sigma}^Q(t)\tilde{\mathbf{Q}}(t). \tag{6.10}$$

It follows from (6.9) and (6.10) that for the exact solutions of problems with superimposed rotation, the following equality should hold:

$$\boldsymbol{\sigma}(t) = \bar{\boldsymbol{\sigma}}^Q(t). \tag{6.11}$$

However, in the numerical integration of CRs for hypoelasticity, this equality can be violated due to errors caused by an inaccurate reproduction of superimposed rotations. Nevertheless, for good algorithms for this integration, equality (6.11) should be satisfied with high accuracy.

Since one of the main kinematic quantities describing the deformation of hypoelastic solids is the stretching tensor $\mathbf{d}$, the accuracy of approximations of the incremental strain determines the accuracy of integrating CRs for hypoelasticity. We estimate the accuracy of these approximations in the investigated problems of homogeneous deformation of samples without superimposed rotations.

For the simple extension problem, the kinematics is in fact one-dimensional and the stretching tensor has a single non-zero component d_{11}. Denoting this component as $d \equiv d_{11}$, we find the accurate expression for the incremental strain e_{incr} and compare it with the expressions for approximate incremental strains presented in Table 4.1.

The equality $d = \dot{F}F^{-1} = \dot{\lambda}_1/\lambda_1 = \overline{\ln \lambda_1}$ leads to the following expression for the incremental strain:

$$e_{\text{incr}} \equiv \int_t^{t+\Delta t} d\, d\tau = \ln \frac{^{t+\Delta t}\lambda_1}{^t\lambda_1}. \tag{6.12}$$

We introduce the notation

$$z \equiv \frac{{}^{t+\Delta t}\lambda_1}{{}^{t}\lambda_1} = \frac{1 + k_1(t + \Delta t)}{1 + k_1 t}. \tag{6.13}$$

In view of notation (6.13), expression (6.12) for the accurate value of the incremental strain can be rewritten as

$$e_{\text{incr}} = \ln z.$$

Note that the following equality holds:

$$F_r = {}^{t+\Delta t}F^t F^{-1} = \frac{{}^{t+\Delta t}\lambda_1}{{}^{t}\lambda_1} = z. \tag{6.14}$$

Taking into account equality (6.14), from (4.19) we obtain

$$^{1/2}H = 2\frac{z-1}{z+1}. \tag{6.15}$$

Then from $(4.21)_1$ and (6.15), we have

$$\Delta t d^m = 2\frac{z-1}{z+1}.$$

Using the Taylor formula, we obtain the power approximation of the function $2(z-1)/(z+1)$ in the vicinity of the point $z = 1$

$$2\frac{z-1}{z+1} = z - 1 - \frac{1}{2}(z-1)^2 + \frac{1}{4}(z-1)^3 + o(z-1)^4. \tag{6.16}$$

The power approximation of the function $\ln z$ in the vicinity of the point $z = 1$ is

$$\ln z = z - 1 - \frac{1}{2}(z-1)^2 + \frac{1}{3}(z-1)^3 + o(z-1)^4. \tag{6.17}$$

Comparison of the expressions on the r.h.s. of (6.16) and (6.17) shows that the midpoint approximate incremental strain $\Delta t d^a_{1/2}$ approximates the exact value of the incremental strain e_{incr} up to third-order terms.

From the expressions for incremental strains in Table 4.1 and the expressions for their non-incremental counterparts in Table 2.1, it follows that the approximate incremental strains $\Delta t d_1 = \Delta t \bar{d}_1$ and $\Delta t d_2 = \Delta t \bar{d}_2$ approximate the exact value of the incremental strain e_{incr} up to the second-order terms, the approximate incremental strains $\Delta t d_4 = \Delta t \bar{d}_4$, $\Delta t d_N = \Delta t \bar{d}_N$, $\Delta t d_W = \Delta t \bar{d}_W$, and $\Delta t d_R = \Delta t \bar{d}_R$ approximate it up to third-order terms, and the quantities $\Delta t d_3 = \Delta t \bar{d}_3$ correspond to the exact value $\ln z$ of the incremental strain. Note that these orders of approximations correspond to the orders of accuracy presented in Table 4.1.

Since, for the simple extension problem, $^{\Delta t}U = {}^{\Delta t}\tilde{V} = U_r = V_r$ (see Chap. 3), all orders of approximations in Table 5.1 correspond to the orders to which the approximate incremental strains in this table approximate the incremental strains $E_{incr} = \tilde{e}_{incr} = e_{incr}$.

Further, for the simple shear problem, we confine ourselves to 2D analysis. In this case, the following expression for the stretching tensor (see, e.g., Eq. $(41)_1$ in [5]) holds:

$$\mathbf{d} = \frac{\dot{\gamma}}{2} \begin{bmatrix} 0 & 1 \\ 1 & 0 \end{bmatrix}.$$

Since $\dot{\gamma} = k_2$, we obtain the exact value of the incremental strain tensor for the simple shear problem:

$$\mathbf{e}_{incr} \left(\equiv \int_t^{t+\Delta t} \mathbf{d} \, d\tau \right) = \frac{\Delta t k_2}{2} \begin{bmatrix} 0 & 1 \\ 1 & 0 \end{bmatrix}. \tag{6.18}$$

Note that $\text{tr}(\mathbf{e}_{incr}) = 0$, which agrees with the isochoric deformation condition for the simple shear problem.

For the simple shear problem, the tensors $\mathbf{F}$ and $\mathbf{F}^{-1}$ can be written as follows (see, e.g., Sect. 6 in [12]):

$$\mathbf{F} = \begin{bmatrix} 1 & \gamma \\ 0 & 1 \end{bmatrix}, \quad \mathbf{F}^{-1} = \begin{bmatrix} 1 & -\gamma \\ 0 & 1 \end{bmatrix}. \tag{6.19}$$

Using equalities $(3.4)_1$, (3.10), and (6.19), we obtain

$$\mathbf{F}_r = \begin{bmatrix} 1 & \Delta t k_2 \\ 0 & 1 \end{bmatrix}. \tag{6.20}$$

From (4.19) and (6.20) we obtain the equality

$$^{1/2}\mathbf{H} = \begin{bmatrix} 0 & \Delta t k_2 \\ 0 & 0 \end{bmatrix}. \tag{6.21}$$

Using equalities $(4.21)_1$ and (6.21), we can write the approximate incremental strain tensor $\Delta t \mathbf{d}^m$ as

$$\Delta t \mathbf{d}^m = \frac{\Delta t k_2}{2} \begin{bmatrix} 0 & 1 \\ 1 & 0 \end{bmatrix}. \tag{6.22}$$

Comparing Eqs. (6.18) and (6.22), we conclude that for the simple shear problem, $\Delta t \mathbf{d}^m = \mathbf{e}_{incr}$, i.e., the midpoint algorithm for this problem leads to the accurate determination of the incremental strain tensor! This conclusion agrees with the similar conclusions drawn in [13, 14].

Using equality (6.20) and the expressions for approximate incremental strains presented from Table 4.1, we obtain

$$\Delta t\mathbf{d}_1 = \frac{\Delta tk_2}{2}\begin{bmatrix} \Delta tk_2 & 1 \\ 1 & 0 \end{bmatrix}, \quad \Delta t\mathbf{d}_2 = \frac{\Delta tk_2}{2}\begin{bmatrix} 0 & 1 \\ 1 & -\Delta tk_2 \end{bmatrix},$$

$$\Delta t\mathbf{d}_4 = \frac{\Delta tk_2}{2}\begin{bmatrix} \Delta tk_2/2 & 1 \\ 1 & -\Delta tk_2/2 \end{bmatrix}. \quad (6.23)$$

In addition, keeping terms up to third order, we obtain the following approximate incremental strain tensor:

$$\Delta t\mathbf{d}_3 = \frac{\Delta tk_2}{2}\left[1 - \frac{(\Delta tk_2)^2}{6}\right]\begin{bmatrix} \Delta tk_2/2 & 1 \\ 1 & -\Delta tk_2/2 \end{bmatrix}. \quad (6.24)$$

Comparison of the expressions for approximate incremental strain tensors in (6.23) and (6.24) with the exact expression (6.18) shows that all these approximate incremental strain tensors approximate the tensor $\mathbf{d}_{incr}$ only up to second-order terms. We conclude that for the simple shear problem, the incremental logarithmic tensor has no advantages over the other approximated incremental strain tensors. This is due to the fact that for the simple shear problem, the second term on the r.h.s. of the expression for the stretching tensor in $(2.13)_2$ plays a significant role. Comparing the expressions for approximate incremental strain tensors in (6.23) and (6.24), we note that for this problem, the tensors $\Delta t\mathbf{d}_1$ and $\Delta t\mathbf{d}_2$ do not maintain the incremental isochoric deformation condition (tr($\Delta t\mathbf{d}_i$) $\neq 0$, $i = 1, 2$), and the tensors $\Delta t\mathbf{d}_3$ and $\Delta t\mathbf{d}_4$ maintain this isochoric condition (tr($\Delta t\mathbf{d}_i$) $= 0$, $i = 3, 4$). We do not analyze the performances of the proposed approximations for the absolutely objective algorithms for the simple shear problem; apparently, these performances are consistent to the performances of strong incrementally objective algorithms.

Summarizing the analysis of the incremental strain tensor approximations considered in this book, we conclude that the Mooney incremental strain tensors proposed in this book is the best choice for approximating incremental strain tensors both for the strong incrementally objective algorithm and for the absolutely objective algorithm. First, these tensors retain all properties of the incremental Hencky strain tensor, and, second, their determination requires fewer operations than the determination of the incremental Hencky strain tensors.

In Sects. 6.3–6.5, the accuracy of the approximate incremental strain tensors considered will be checked by solving the simple extension and simple shear problems for all hypoelastic material models considered in this book.

6.2 Background for Computer Simulations

Computer simulations of simple extension and simple shear deformations of a 1 m × 1 m square sample were performed using the homemade FE code to implement all material models considered in this paper. We use the material parameters (Young's modulus E and Poisson's ratio ν)

$$E = 3.37\,\text{MPa}, \quad \nu = 0.45,$$

to determine the Lamé parameters from the expressions (see, e.g., Table 5.1 in [15])

$$\lambda = \frac{E\nu}{(1+\nu)(1-2\nu)} = 10.46\,\text{MPa}, \quad \mu = \frac{E}{2(1+\nu)} = 1.16\,\text{MPa}.$$

We adopt the following values for the parameters k_1, k_2, and k_3 (see Sect. 6.1):

$$k_1 = k_2 = k_3 = 10\,\text{s}^{-1}.$$

The time integration interval is 1 s.

The convergence of numerical solutions to the exact solutions is checked using a set of time steps with a fourfold decrease in the integration time step, starting at a value $\Delta t = 1/20$ sec and ending at a value $\Delta t = 1/1280$ s (i.e., the time integration interval of 1 sec is sequentially divided into 20, 80, 320, and 1280 time steps). For each of these values of the time step, the increment of the longitudinal engineering strain $\epsilon \equiv \lambda - 1$ in the simple extension problem, the increment of the shear parameter γ for the simple shear problem, and the increment of the angle of superimposed rotation θ (if present) are given in Table 6.1.

To evaluate the performance of a certain considered algorithm, we determine the relative errors e_{ij} in the Cauchy stresses σ_{ij} (i, j run 1, 2) using the slightly modified expression given on p. 280 in Drozdov [16]:

$$e_{ij}^2 \equiv \frac{1}{M} \Sigma_{m=1}^{M} \left[\frac{\sigma_{ij}^{\text{num}}(t_m) - \sigma_{ij}^{\text{exact}}(t_m)}{\sigma_{ij\,\text{max}}^{\text{num}}} \right]^2 \quad (i, j \text{ run 1, 2}).$$

Table 6.1 Values of increments of the longitudinal engineering strain ϵ for the simple extension problem, the shear parameter γ for the simple shear problem, and the angle of superimposed rotation θ versus the number of time steps for integrating CRs for hypoelastic models in a time interval of 1 s

Number of time steps M	Power k	$\Delta\epsilon$	$\Delta\gamma$	$\Delta\theta$ (in deg.)
20	0	1/2	1/2	180.0
80	1	1/8	1/8	45.0
320	2	1/32	1/32	11.25
1280	3	1/128	1/128	2.8125

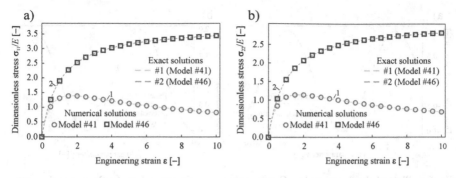

Fig. 6.2 Plots of the dimensionless Cauchy stresses σ_{11}/E (**a**) and σ_{22}/E (**b**) versus engineering strain ϵ for the simple extension problem when integrating CRs for material models #41 and #46 using the midpoint algorithms with 20 time steps

Hereinafter, $\sigma_{ij}^{\text{num}}(t_m)^2$ and $\sigma_{ij}^{\text{exact}}(t_m)$ (i, j run 1, 2) are the values of the Cauchy stress tensor components obtained by numerical integration of the considered CRs and their exact values, respectively, determined at times t_m ($m = 1, ...M$) (M is the number of time steps). The quantity $\sigma_{ij\,\text{max}}^{\text{num}}$ is the maximum absolute value of σ_{ij}^{num}, i.e.,

$$\sigma_{ij\,\text{max}}^{\text{num}} \equiv \max_{t_m=1/M,...,1\ \text{s}} |\sigma_{ij}^{\text{num}}(t_m)| \quad (i,\ j\ \text{run 1, 2}).$$

Note that the numbers of time steps used in the solutions of problems can be represented as $M = 20 \cdot 4^k$. The values of the power k are shown in Table 6.1. These values of the power k will be used to construct plots of errors in stresses versus this power in Sects. 6.3–6.5.

To integrate CRs, we use one four-node element for plane strain with a 2×2 order of Gauss integration. Since the deformations of the square sample are homogeneous, the obtained solutions for stresses are similar at all integration points.

6.3 Simulations Using Weak Incrementally Objective Algorithms

In this section, we compare two weak incrementally objective algorithms (the H-W and R-A ones) and establish the influence of the SRBMs on the accuracy of determining stresses for the considered problems of uniform deformation of hypoelastic bodies.

---

[2] In the solutions of problems with superimposed rotations, the notations $\sigma_{ij}^{\text{num}}(t_m)$ are used to refer to the components of the back rotated Cauchy stress tensor $\bar{\sigma}_{ij}^{Q}(t_m)$.

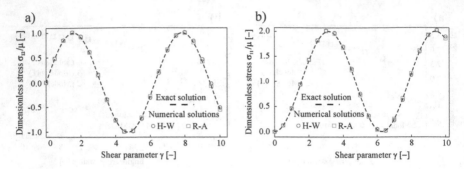

Fig. 6.3 Plots of the dimensionless Cauchy stresses σ_{12}/μ (**a**) and σ_{11}/μ (**b**) versus the shear parameter γ for the simple shear problem when integrating CRs for material model #41 using midpoint algorithms with 20 time steps

6.3.1 Solutions of Problems without Superimposed Rotations

Consider the solutions of the simple extension problem. Since the solutions of this problem for material models ##41–43 are similar, we solve this problem only for material models #41 and #46. Since there are no rotations of material fibers in this problem, the solutions of this problem using the H–W and R–A algorithms are indistinguishable from each other. Figure 6.2 shows the plots of the dimensionless Cauchy stresses σ_{11}/E and σ_{22}/E versus engineering strain ϵ obtained by integrating CRs for material models #41 and #46 using the midpoint algorithm with 20 time steps.[3] Since the obtained values of the Cauchy stress tensor components are fairly close to their exact values, further refinement of the time step was not performed.

Consider the solutions of the simple shear problem. Figures 6.3 and 6.4 show plots of the dimensionless Cauchy stresses σ_{12}/μ and σ_{11}/μ versus the shear parameter γ obtained by integrating CRs for material models #41 and #46 (Fig. 6.3)[4] and #42 and #43 (Fig. 6.4) using 20 time steps. We see that both the H–W and R–A algorithms are able to fairly accurately reproduce the exact solution in integrating CRs for these material models using 20 time steps, but the accuracy of reproduction is higher for the R–A algorithm than for the H–W algorithm, which is consistent with the theoretical propositions presented in Sect. 4.2. As in the solution of the simple extension problem, the obtained values of the Cauchy stress tensor components are fairly close to their exact values, so that further refinement of the time step was not performed.

---

[3] Note that dependence of the component σ_{22}/E on engineering strain ϵ for the simple extension problem is determined exactly for material models ##41–43 ($\sigma_{22} = \lambda \ln J/J = \lambda \ln \lambda_1/\lambda_1$) and material model #46 ($\sigma_{22} = \lambda(J - 1)/J = \lambda(\lambda_1 - 1)/\lambda_1$).

[4] Since the solutions of the simple shear problem for material models #41 and #46 coincide, Fig. 6.3 shows the solution of this problem only for material model #41.

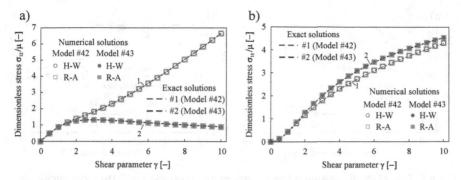

Fig. 6.4 Plots of the dimensionless Cauchy stresses σ_{12}/μ (**a**) and σ_{11}/μ (**b**) versus the shear parameter γ for the simple shear problem when integrating CRs for material models #42 and #43 using midpoint algorithms with 20 time steps

6.3.2 Solutions of Problems with Superimposed Rotations

Since, in simulations of deformations for both the simple extension and simple shear problems, the influence of superimposed rotations is qualitatively similar for the material models considered, we confine our study to material model #41. As might be expected (cf., [15, 17–21]), the presence of rotations superimposed on the main deformation greatly deteriorates the quality of approximation of the desired quantities for weak incrementally objective algorithms. Computer simulations of simple extension for sequences of refined time steps are presented in Fig. 6.5. Plots of the errors e_{11} and e_{22} in the Cauchy stresses σ_{11} and σ_{22} versus the power k are shown in Fig. 6.6. The solutions of this problem show the advantage of the R–A algorithm over the H–W algorithm. In particular, in the solution of this problem using the R–A algorithm, values of the stress tensor components close to the exact solution are obtained after 320 time steps, whereas the H–W algorithm yields the same values after 1280 time steps.

Computer simulations of simple shear using 80 and 320 time steps are presented in Figs. 6.7 and 6.8, respectively. Plots of the errors e_{12} and e_{11} in the Cauchy stresses σ_{12} and σ_{11} versus the power k are shown in Fig. 6.9. We see that the solution of the simple shear problem obtained using the H–W algorithm is closer to the exact solution than that obtained using the R–A algorithm! This conclusion is inconsistent with theoretical estimates of the performances of these algorithms (see, e.g., [22]) and contradicts the conclusion drawn above when solving the simple extension problem with superimposed rotation. Note that in computer simulations with 1280 time steps, both algorithms provide values of the stress tensor components close to the exact values.

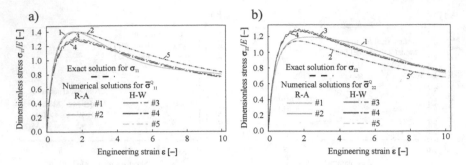

Fig. 6.5 Plots of the dimensionless Cauchy stresses σ_{11}/E (**a**) and σ_{22}/E (**b**) versus engineering strain ϵ when integrating CRs for material model #41 using midpoint algorithms for the simple extension problem with superimposed rotation: the dashed curves correspond to the exact solution; the solid curves #1 and #2 correspond to computer simulations using the R–A algorithm with 80 and 320 time steps, respectively; the dashed-dotted curves ##3–5 correspond to computer simulations using the H–W algorithm with 80, 320, and 1280 time steps, respectively

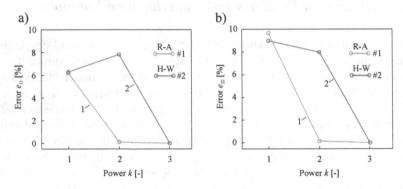

Fig. 6.6 Plots of the errors e_{11} (**a**) and e_{22} (**b**) in the Cauchy stresses σ_{11} and σ_{22} versus the power k when integrating CRs for material model #41 using midpoint algorithms for the simple extension problem with superimposed rotation

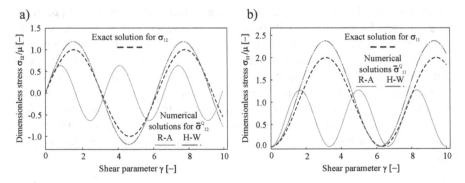

Fig. 6.7 Plots of the dimensionless Cauchy stresses σ_{12}/μ (**a**) and σ_{11}/μ (**b**) versus the shear parameter γ when integrating CRs for material model #41 using midpoint algorithms with 80 time steps for the simple shear problem with superimposed rotation

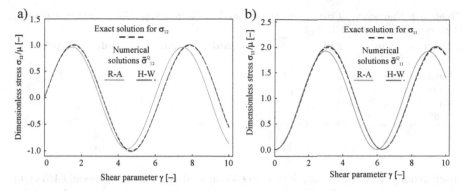

Fig. 6.8 Plots of the dimensionless Cauchy stresses σ_{12}/μ (**a**) and σ_{11}/μ (**b**) versus the shear parameter γ when integrating CRs for material model #41 using midpoint algorithms with 320 time steps for the simple shear problem with superimposed rotation

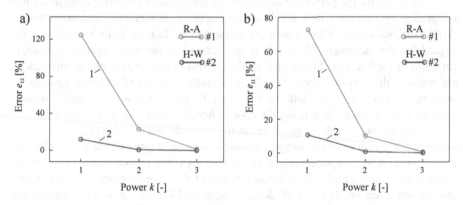

Fig. 6.9 Plots of the errors e_{12} (**a**) and e_{11} (**b**) in the Cauchy stresses σ_{12} and σ_{11} versus the power k when integrating CRs for material model #41 using midpoint algorithms for the simple shear problem with superimposed rotation

6.4 Simulations Using Strong Incrementally Objective Algorithms

Numerical experiments show that, as might be expected, rotations superimposed on simple extension and simple shear do not affect the results of computer simulations using strong incrementally objective algorithms, i.e., equality (6.11) is satisfied with the maximum accuracy possible in the presence of errors in representations of numbers with double precision. Therefore, in this section we do not specify whether or not rotations are superimposed on the basic deformations for computer simulations of simple extension and simple shear. In addition, both versions of strong incrementally objective algorithms (incremental Eulerian or Lagrangian objectivity in the determination of the Kirchhoff stress tensor from expressions (4.66) and (4.67), respectively)

Table 6.2 Numbering of the incrementally Eulerian-objective approximate incremental strain tensors $\Delta t \mathbf{d}^a$ presented in Table 4.1

Tensor $\Delta t \mathbf{d}^a$	$\Delta t \mathbf{d}_1$	$\Delta t \mathbf{d}_2$	$\Delta t \mathbf{d}_3$	$\Delta t \mathbf{d}_4$	$\Delta t \mathbf{d}_N$	$\Delta t \mathbf{d}_W$	$\Delta t \mathbf{d}_R$
Number of the tensor $\Delta t \mathbf{d}^a$	1	2	3	4	5	6	7

used to integrate CRs for material model #41 (as well as for material model #46) yield the same values for the Kirchhoff stress tensor $^{t+\Delta t}\boldsymbol{\tau}$. Therefore, here we only give the results of simulations using the strong incrementally Eulerian-objective algorithm to integrate CRs for models #41 and #46.

A key factor for the performance of strong incrementally objective algorithms for integrating hypoelastic CRs is the form of the approximate incremental strain tensor $\Delta t \mathbf{d}^a$. Various forms of this tensor are presented in Table 4.1. In Sect. 4.3.2, we argued that the incremental Mooney strain tensor $\mathbf{e}_r^M$ is the most appropriate choice (in terms of the balance of price and quality) of this tensor. We check the theoretical estimates of the performances of the incrementally Eulerian-objective approximate incremental strain tensors $\Delta t \mathbf{d}^a$ in Table 4.1 by computer simulations of simple extension and simple shear using CRs for material model #41. For ease of further references, we number these quantities in the order shown in Table 6.2.

Consider solutions of the simple extension problem. Figure 6.10a shows plots of the dimensionless component σ_{11}/E of the Cauchy stress tensor versus engineering strain ϵ when integrating CRs for material model #41 using the approximate incremental strain tensors ##1–7 with 20 time steps and Fig. 6.10b shows plots of the errors e_{11} in the Cauchy stress σ_{11} versus the power k. Similar plots for the component σ_{22}/E are not given here since this component for the simple extension problem is determined exactly ($\sigma_{22} = \lambda \ln J/J = \lambda \ln \lambda_1/\lambda_1$) regardless of which of the quantities $\Delta t \mathbf{d}^a$ is used to simulate deformations. The orders of accuracy of the plots in Fig. 6.10 in relation to the exact solution correspond to the orders of approximation of the incremental strain tensor $\int_t^{t+\Delta t} \mathbf{d}\, d\tau$ given in the last column of Table 4.1. In particular, approximation #3 leads to an accurate determination of the incremental strain tensor and hence to an accurate determination of the Cauchy stress tensor component σ_{11}; approximations #1 and #2 lead to the least accurate determination of the quantity σ_{11}, and approximations ##4–7 lead to a more accurate determination of this quantity; however, among the latter approximations, approximation #7 proposed by Rashid [17] is the least accurate.

Consider solutions of the simple shear problem. Figure 6.11 shows plots of the dimensionless Cauchy stresses σ_{12}/μ and σ_{11}/μ versus the shear parameter γ when integrating CRs for material model #41 using 20 time steps. The curves presented in this figure show that the simulations using approximations #3, #4, and #6 adequately reproduce the exact solution for the stress tensor components considered. The curves

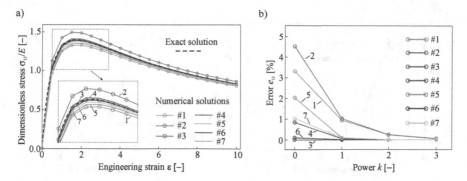

Fig. 6.10 Plots of (**a**) the dimensionless Cauchy stress σ_{11}/E versus engineering strain ϵ for the simple extension problem when integrating CRs for material model #41 using the strong incrementally objective algorithm with 20 time steps: the dashed curve corresponds to the exact solution, and the solid curves ##1–7 correspond to simulations using the approximate incremental strain tensors ##1–7, respectively; (**b**) the errors e_{11} in the Cauchy stress σ_{11} versus the power k

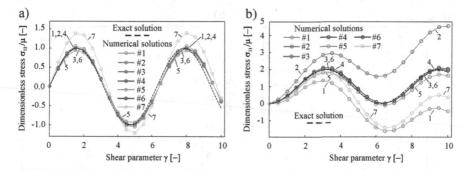

Fig. 6.11 Plots of the dimensionless Cauchy stresses σ_{12}/μ (**a**) and σ_{11}/μ (**b**) versus the shear parameter γ for the simple shear problem when integrating CRs for material model #41 using the strong incrementally objective algorithm with 20 time steps: the dashed curves correspond to the exact solution, and the solid curves ##1–7 to simulations using the approximate incremental strain tensors ##1–7, respectively

of the solutions for approximation #5 are fairly close to the curves of the solutions for these approximations. However, the curves of the solutions for approximations #1, #2, and #7 correlate poorly with the exact solution curves; this is especially pronounced for the curves of σ_{11}/μ versus γ. Since we suggest that approximation #4 is the optimal choice for computer simulations, in Fig. 6.12, the plots of the dimensionless Cauchy stresses σ_{12}/μ and σ_{11}/μ versus the shear parameter γ for integration of CRs for material model #41 with 20 time steps are given only for approximations #3 and #4. The curves show that the stress tensor components obtained using these approximations are in fairly good agreement with each other and with exact solution of the simple shear problem.

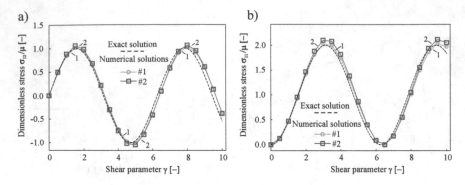

Fig. 6.12 Plots of the dimensionless Cauchy stresses Cauchy stresses σ_{12}/μ (**a**) and σ_{11}/μ (**b**) versus the shear parameter γ for the simple shear problem when integrating CRs for material model #41 using the strong incrementally objective algorithm with 20 time steps: the dashed curves correspond to the exact solution, and the solid curves #1 and #2 to simulations using the approximate incremental strain tensors #3 and #4, respectively

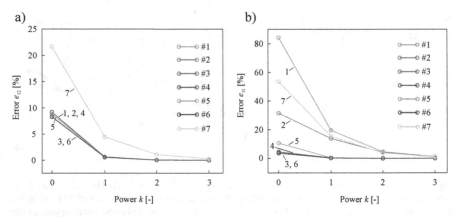

Fig. 6.13 Plots of the errors e_{12} (**a**) and e_{11} (**b**) in the Cauchy stresses σ_{12} and σ_{11} versus the power k for the simple shear problem when integrating CRs for material model #41 using the strong incrementally objective algorithm: the curves ##1–7 correspond to simulations using the approximate incremental strain tensors ##1–7, respectively

Plots of the errors e_{12} and e_{11} in the Cauchy stresses σ_{12} and σ_{11} versus the power k are shown in Fig. 6.13. The results of computer simulations using 80, 320, and 1280 time steps show the stress tensor components obtained using approximations #3, #4, #5, and #6 are fairly accurately reproduced in the computer simulations with 80 time steps. The same quantities are well reproduced using approximation #7 with 1280 time steps, whereas computer simulations using approximations #1 and #2 reproduces the dimensionless Cauchy stress σ_{11}/μ with insufficient accuracy even when using 1280 time steps.

Thus, the theoretical estimates of the performances of the incrementally Eulerian-objective approximate incremental strain tensors $\Delta t \mathbf{d}^a$ given in Table 4.1 are con-

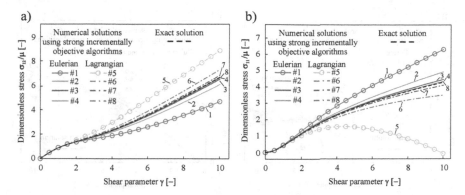

Fig. 6.14 Plots of the dimensionless Cauchy stresses σ_{12}/μ (**a**) and σ_{11}/μ (**b**) versus the shear parameter γ for the simple shear problem when integrating CRs for material model #42 using the strong incrementally objective algorithm with the approximate incremental strain tensor #4: the dashed curves correspond to the exact solution, and the solid curves ##1–4 (and the dashed-dotted curves ##5–8) to simulations using the strong incrementally Eulerian-objective algorithm (and the strong incrementally Lagrangian-objective algorithm) with 20, 80, 320, and 1280 time steps, respectively

firmed by computer simulations of simple extension and simple shear using CRs for material model #41; namely, the tensor $\Delta t \mathbf{d}_4 = \mathbf{e}_r^M$ is the best choice among the approximate incremental strain tensors $\Delta t \mathbf{d}^a$ shown in Table 4.1. A similar conclusion is also true for computer simulations (not presented here) of simple extension and simple shear using CRs for material models ##42, 43, 46. For further computer simulations of simple extension and simple shear for these material models, we use only the approximate incremental strain tensor $\Delta t \mathbf{d}_4$.

Figures 6.14 and 6.15 show plots of the dimensionless Cauchy stresses σ_{12}/μ and σ_{11}/μ versus the shear parameter γ obtained for the simple shear problem when integrating CRs for material models #42 (Fig. 6.14) and #43 (Fig. 6.15) using 20, 80, 320, and 1280 time steps, respectively. Here we use both versions (Lagrangian- and Eulerian-objective) of strong incrementally objective algorithms. Plots of the errors e_{12} and e_{11} in the Cauchy stresses σ_{12} and σ_{11} versus the power k for material models #42 and #43 are shown in Fig. 6.16. It can be seen from the curves in Figs. 6.14, 6.15 and 6.16 that for the approximate incremental strain tensor #4, acceptable values of the Cauchy stress components are obtained using only 320 time steps, and when using 1280 time steps, the numerical solution curve merge (almost merge) with the exact solution curve. These results differ greatly from the results of the solution obtained by integrating CRs for material model #41, where acceptable values of the Cauchy stress components are obtained using only 20 time steps (see Figs. 6.11, 6.12 and 6.13), and when using 80 time steps, the numerical solution curves are indistinguishable from the exact solution curves. In addition, both versions of the strong incrementally objective algorithms yield close values of the Cauchy stresses only in computer simulations using 1280 time steps.

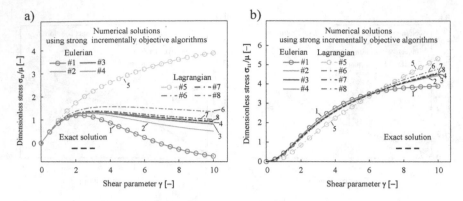

Fig. 6.15 Plots of the dimensionless Cauchy stresses σ_{12}/μ (**a**) and σ_{11}/μ (**b**) versus shear parameter γ for the simple shear problem when integrating CRs for material model #43 using the strong incrementally objective algorithm with the approximate incremental strain tensor #4: the dashed curves correspond to the exact solution, and the solid curves ##1–4 (and the dashed-dotted curves ##5–8) to simulations using the strong incrementally Eulerian-objective algorithm (and the strong incrementally Lagrangian-objective algorithm) with 20, 80, 320, and 1280 time steps, respectively

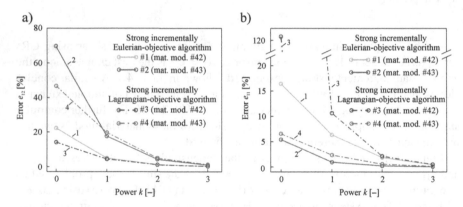

Fig. 6.16 Plots of the errors e_{12} (**a**) and e_{11} (**b**) in the Cauchy stresses σ_{12} and σ_{11} versus the power k for the simple shear problem when integrating CRs for material models #42 and #43 using the strong incrementally objective algorithms with the approximate incremental strain tensor #4: the curves ##1, 2 correspond to simulations using the incrementally Eulerian-objective algorithm and the curves ##3, 4 correspond to simulations using the incrementally Lagrangian-objective algorithm for material models #42 and #43, respectively

Consider the solutions of the simple extension problem when integrating CRs for material model #46 using the approximate incremental strain tensor #4 with 20 and 80 time steps. Figure 6.17a shows plots of the dimensionless component σ_{11}/E of the Cauchy stress tensor versus engineering strain ϵ, and Fig. 6.17b shows plots of the errors e_{11} in the Cauchy stress σ_{11} versus the power k. It can be seen from the curves presented in Fig. 6.17 that acceptable values of the Cauchy stress components are obtained using only 20 time steps, and when using 80 time steps,

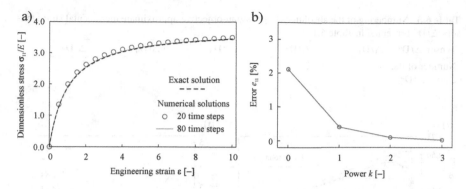

Fig. 6.17 Plots of (**a**) the dimensionless component σ_{11}/E of Cauchy stress tensor versus engineering strain ϵ in the simple extension problem when integrating CRs for material model #46 using the strong incrementally objective algorithm with the approximate incremental strain tensor #4; (**b**) the errors e_{11} in the Cauchy stress σ_{11} versus the power k

the numerical solution curve merges with the exact solution curve. We do not give similar dependencies for the component σ_{22}/E since this component for the simple extension problem is determined exactly ($\sigma_{22} = \lambda(J - 1)/J = \lambda(\lambda_1 - 1)/\lambda_1)$).

The solution of the simple shear problem obtained by integrating CRs for material model #46 coincides with the above solution for the same problem obtained by integrating CRs for material model #41.

6.5   Simulations Using Absolutely Objective Algorithms

The results of the numerical experiments show that as in the case of strong incrementally objective algorithms, rotations superimposed on simple extension and simple shear do not affect the results of computer simulations using absolutely objective algorithms. In the determination of the rotated and standard Kirchhoff stress tensor from expressions (5.13) and (5.27), respectively, both versions of the absolutely objective algorithm (Lagrangian or Eulerian) yield the same values of the Kirchhoff stress tensor $^{t+\Delta t}\tau$. Nevertheless, the Lagrangian version of the absolutely objective algorithm requires fewer arithmetic operations compared to the Eulerian version, therefore, here we present the results of simulations using the absolutely Lagrangian-objective algorithm.

The form of the approximate incremental strain tensor $\Delta t \mathbf{D}^a$ is a key factor for the performance of absolutely objective algorithms for integrating hypoelastic CRs. We check the theoretical estimates of the performances of the absolutely Lagrangian-objective approximate incremental strain tensors $\Delta t \mathbf{D}^a$ presented in Table 5.1 by computer simulations of simple extension and simple shear using CRs for material model #41. For convenience of further references, we number these quantities in the order shown in Table 6.3.

Table 6.3 Numbering of the absolutely Lagrangian-objective approximate incremental strain tensors $\Delta t \mathbf{D}^a$ presented in Table 5.1

Tensor $\Delta t \mathbf{D}^a$	$\Delta t \mathbf{D}_1$	$\Delta t \mathbf{D}_2$	$\Delta t \mathbf{D}_3$	$\Delta t \mathbf{D}_4$	$\Delta t \mathbf{D}_5$
Number of the tensor $\Delta t \mathbf{D}^a$	1	2	3	4	5

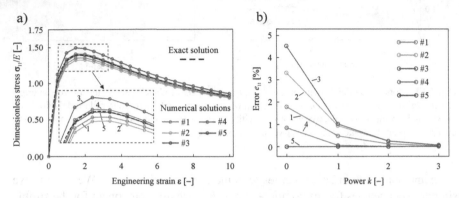

Fig. 6.18 Plots of (**a**) the dimensionless component σ_{11}/E of the Cauchy stress tensor versus engineering strain ϵ for the simple extension problem when integrating CRs for material model #41 using the absolutely objective algorithm with 20 time steps: the dashed curve corresponds to the exact solution, and the solid curves ##1–5 to simulations using the approximate incremental strain tensors ##1–5, respectively; (**b**) the errors e_{11} in the Cauchy stress σ_{11} versus the power k

Consider solutions of the simple extension problem. Figure 6.18a shows plots of the dimensionless component σ_{11}/E of the Cauchy stress tensor versus engineering strain ϵ obtained using the approximate incremental strain tensors ##1–5 with 20 time steps to integrate CRs for material model #41, and Fig. 6.18b shows plots of the errors e_{11} in the Cauchy stress σ_{11} versus the power k. As in the case of integration of CRs for this material model using the strong incrementally objective algorithm, the similar plot for the component σ_{22}/E is not shown here since this component for the simple extension problem is determined exactly $\sigma_{22} = \lambda \ln \lambda_1/\lambda_1$ regardless of which of the quantities $\Delta t \mathbf{D}^a$ is used to simulate deformations. The orders of accuracy of the plots in Fig. 6.18 in relation to the exact solution correspond to the orders of approximations of the incremental strain tensor $\int_t^{t+\Delta t} \mathbf{D} \, d\tau$ given in the last column of Table 5.1. In particular, approximation #5 leads to an accurate determination of the incremental strain tensor and hence to an accurate determination of the Cauchy stress tensor component σ_{11}. Approximations #2 and #3 lead to the least accurate determination of the quantity σ_{11}, and approximations #1 and #4 lead to a more accurate determination of this quantity, but approximation #4 is more accurate than approximation #1. The numerical solution curves (not shown here) obtained by integrating CRs for material model #41 using the approximate incremental strain tensors ##1–5 with 80 and 320 time steps, respectively, are closer to the exact solution curve (see also Fig. 6.18b), but qualitative behavior of these curves does not change.

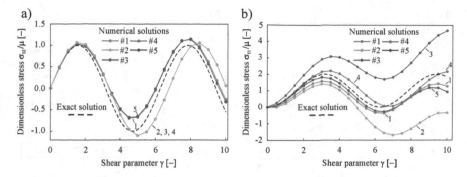

Fig. 6.19 Plots of the dimensionless Cauchy stresses σ_{12}/μ (**a**) and σ_{11}/μ (**b**) versus the shear parameter γ for the simple shear problem when integrating CRs for material model #41 using the absolutely objective algorithm with 20 time steps: the dashed curves correspond to the exact solution, and the solid curves ##1–5 to simulations using the approximate incremental strain tensors ##1–5, respectively

The solutions of the same problem when integrating CRs for material model #41 using 1280 time steps lead to curves of the dimensionless component σ_{11}/E of the Cauchy stress tensor versus engineering strain ϵ that almost merge with each other and with the exact solution curve of the simple extension problem.

Consider solutions of the simple shear problem. Figure 6.19 shows curves of the dimensionless Cauchy stresses σ_{12}/μ and σ_{11}/μ versus the shear parameter γ when integrating CRs for material model #41 using 20 time steps. The curves presented in this figure show that approximations #1, #4, and #5 fairly well reproduce the exact solution for the stress tensor components considered. However, the curves of the solutions for approximations #2 and #3 correlate poorly with the exact solution curves; this is especially pronounced for the curves of σ_{11}/μ versus γ.

Plots of the errors e_{12} and e_{11} in the Cauchy stresses σ_{12} and σ_{11} versus the power k are shown in Fig. 6.20. The results of computer simulations for the same problem using 80, 320, and 1280 time steps show that stress tensor components obtained using approximations #1, #4, and #5 are fairly accurately reproduced in computer simulations with 320 time steps, and computer simulations using approximations #2 and #3 insufficient accurately reproduce the dimensionless Cauchy stress σ_{11}/μ even when using 1280 time steps.

Thus, the theoretical estimates in Table 5.1 for the performances of the absolutely Lagrangian-objective approximate incremental strain tensors $\Delta t\mathbf{D}^a$ are confirmed by computer simulations of simple extension and simple shear obtained using CRs for material model #41; namely, the incremental right Mooney strain tensor $\Delta t\mathbf{D}_4$ is the best choice among the approximate incremental strain tensors $\Delta t\mathbf{D}^a$ given in Table 5.1. A similar conclusion is also true for computer simulations (not presented here) of simple extension and simple shear using CRs for material models ##42, 43, 46. For further computer simulations of simple extension and simple shear for these material models, we use only the approximate incremental strain tensor $\Delta t\mathbf{D}_4$.

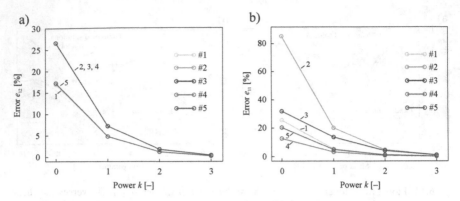

Fig. 6.20 Plots of the errors e_{12} (**a**) and e_{11} (**b**) in the Cauchy stresses σ_{12} and σ_{11} versus the power k for the simple shear problem when integrating CRs for material model #41 using the absolutely objective algorithm: the curves ##1–5 correspond to simulations using the approximate incremental strain tensors ##1–5, respectively

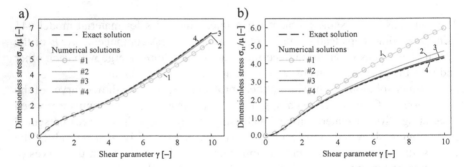

Fig. 6.21 Plots of the dimensionless Cauchy stresses σ_{12}/μ (**a**) and σ_{11}/μ (**b**) versus the shear parameter γ for the simple shear problem when integrating CRs for material model #42 using the absolutely objective algorithm with the approximate incremental strain tensor #4: the dashed curves correspond to the exact solutions, and the solid curves ##1–4 to simulations using 20, 80, 320, and 1280 time steps, respectively

   Figures 6.21 and 6.22 show curves of the dimensionless Cauchy stresses σ_{12}/μ (**a**) and σ_{11}/μ (**b**) versus the shear parameter γ for the simple shear problem when integrating CRs for material models #42 and #43 using the absolutely objective algorithm. It can be seen from the curves that acceptable values of the Cauchy stress components are obtained using only 320 time steps, and when using 1280 time steps, the numerical solution curves merge (almost merge) with the exact solution curves. Plots of the errors e_{12} and e_{11} in the Cauchy stresses σ_{12} and σ_{11} versus the power k for material models #42 and #43 are shown in Fig. 6.23.

   The solution of the simple extension problem obtained by integrating CRs for material model #46 using the absolutely objective algorithm with the approximate incremental strain tensor #4 using 20 and 80 time steps coincides with the solution

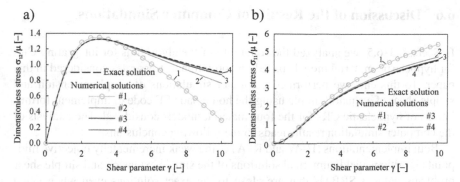

Fig. 6.22 Plots of the dimensionless Cauchy stresses σ_{12}/μ (**a**) and σ_{11}/μ (**b**) versus the shear parameter γ for the simple shear problem when integrating CRs for material model #43 using the absolutely objective algorithm with the approximate incremental strain tensor #4: the dashed curves correspond to the exact solution, and the solid curves ##1–4 to simulations using 20, 80, 320, and 1280 time steps, respectively

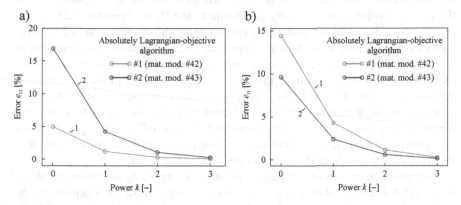

Fig. 6.23 Plots of the errors e_{12} (**a**) and e_{11} (**b**) in the Cauchy stresses σ_{12} and σ_{11} versus the power k for the simple shear problem when integrating CRs for material models #42 and #43 using the absolutely objective algorithm with the approximate incremental strain tensor #4: the curves #1 and #2 correspond to simulations for material models #42 and #43, respectively

of this problem obtained using the strong incrementally objective algorithm (see Fig. 6.17, which shows plots of the dimensionless component σ_{11}/E of the Cauchy stress tensor versus engineering strain ϵ and the errors e_{11} in the Cauchy stress σ_{11} versus the power k).

The solution of the simple shear problem obtained by integrating CRs for material model #46 coincides with the solution of the same problem obtained by integrating CRs for material model #41. In particular, using the approximate Lagrangian incremental strain tensor #4, we obtain the dependencies of the dimensionless Cauchy stresses σ_{12}/μ (*a*) and σ_{11}/μ (*b*) on the shear parameter γ shown by curves #4 in Fig. 6.19.

6.6 Discussion of the Results of Computer Simulations

In Sects. 6.3–6.5, we analyzed the reliability of the algorithms for integrating CRs for hypoelastic material models based on corotational stress rates considered in this paper. To this end, we performed computer simulations of simple extension and simple shear of a square sample using the homemade FE code to implement formulations of hypoelastic CRs for the four material models considered. The analysis of the computer simulation results leads to the following conclusions.

Both implementations (H–W and R–A) of the weak incrementally objective midpoint algorithms yield numerical solutions of the simple extension and simple shear problems without SRBMs that are close to the exact solutions even when using coarse time steps. In the presence of additional SRBMs, the R–A algorithm provides a more accurate solution to the simple extension problem compared to the H–W algorithm, and, conversely, the H–W algorithm provides a more accurate solution to the simple shear problem with SRBMs compared to the R–A algorithm. Nevertheless, in computational practice, it is recommended to use the R–A algorithm since the R–A algorithm is theoretically preferable to the H–W algorithm [22]. Nevertheless, for some problems with some special deformations accompanied by large rotations, both of these algorithms do not provide these solutions for coarse time steps (see Remark 4.3). At the same time, strong incrementally objective and absolutely objective algorithms do not have this disadvantage, although in the case of some deformations without SRBMs, they are less accurate than midpoint algorithms.[5]

For strong incrementally objective and absolutely objective algorithms, as noted above, the incremental Mooney strain tensors $\Delta t\mathbf{d}_4$ and $\Delta t\mathbf{D}_4$ are the optimal choice of approximate incremental strain tensors. These theoretical estimates are confirmed by the results of computer simulations presented in Sects. 6.4 and 6.5.

Comparing the results of computer simulations using strong incrementally objective (Sect. 6.4) and absolutely objective (Sect. 6.5) algorithms, we conclude that the use of strong incrementally objective algorithms is preferred for integrating CRs for material models #41 and #46 (C-models) and the use of absolutely objective algorithms is preferred for integrating CRs for material models #42 and #43 (G-models). This is apparently due to the fact that CRs for C-models are based on the vorticity tensor $\mathbf{w}$, which does not depend on the choice of the reference configuration, and CRs for G-models are based on the spin tensors ω^R and $\omega^{\log}$, which depend on the choice of this configuration.

---

[5] Rodriguez-Ferran et al. [23, 24] have arrived at similar conclusions when comparing solutions of simple shear problems using CRs for a hypoelastic model based on the upper Oldroyd stress rate. They have found that the solution of this problem using a midpoint algorithm has second-order accuracy and the solution of the same problem using a strong incrementally objective algorithm (although these authors do not use the term strong incremental objectivity) has only first-order accuracy compared to the exact solution. Nevertheless, numerical experiments of these authors have shown that for the problem of extension of a block with superimposed rotation latter, the second algorithm yields a more accurate solution than the first algorithm. The present work (as the works of other authors [15, 20, 21, 25, 26]) explains this seemingly paradoxical result of these numerical experiments.

We recommend that developers of commercial FE codes should implement two algorithms for integrating CRs for each of the hypoelasticity models considered; i.e., they should suggest the user by default to use strong incrementally and absolute objective algorithms for models of classical and generalized hypoelasticity, respectively, and the same time, provide the user with the option to apply weak incrementally objective mid-point algorithms that in simulations of some types of deformations provide high-quality integration of CRs for all these material models (see Figs. 6.2, 6.3 and 6.4).

In this work, we deliberately simulated only the simplest types of deformation (simple elongation and simple shear), focusing only on the efficiency and reliability of algorithms for numerical integration of the hypoelasticity models considered and leaving aside the performances of these models themselves. For example, it is known (see Remark 4.6) that in the absence of initial stresses (i.e., $\sigma^0 = \mathbf{0}$ at $t = t_0$), the hypoelastic model #43 (based on the logarithmic stress rate) is the rate form of the Hencky isotropic hyperelastic model. Therefore, in closed single parameter cycles, the use of this material model does not lead to undesirable accumulation of stresses, but for other material models, stress accumulation can occur (see, e.g., [27]). However, hypoelastic material models ##41, 42, 46 are already implemented in existing commercial FE codes (cf., [28–30]) as constituents of CRs for models of additive elaso-plasticity, which encourages the development of improved algorithms for integrating their CRs. Apparently, in the near future, the hypoelasticity model based on the logarithmic stress rate will also be implemented as a constituent of CRs for models of additive elasto-plasticity in commercial FE codes since these models are also models of multiplicative elasto-plasticity (cf., [31, 32]). Therefore, the development of improved algorithms for integrating CRs for hypoelastic model #43 is also an important task of computational mechanics.

References

1. M.A. Crisfield, *Non-linear Finite Element Analysis of Solids and Structures: Vol. 2. Advanced Topics* (Wiley, Chichester, 1997)
2. S.N. Atluri, Comput. Methods Appl. Mech. Eng. **43**, 137 (1984). https://doi.org/10.1016/0045-7825(84)90002-1
3. J.K. Dienes, Acta Mech. **32**, 217 (1979). https://doi.org/10.1007/BF01379008
4. L. Szabó, M. Balla, Int. J. Solids Struct. **25**, 279 (1989). https://doi.org/10.1016/0020-7683(89)90049-8
5. S.N. Korobeynikov, Arch. Appl. Mech. **90**, 313 (2020). https://doi.org/10.1007/s00419-019-01611-3
6. S.N. Korobeynikov, J. Elast. **136**, 159 (2019). https://doi.org/10.1007/s10659-018-9699-9
7. H. Xiao, O.T. Bruhns, A. Meyers, Acta Mech. **124**, 89 (1997). https://doi.org/10.1007/BF01213020
8. H. Xiao, O.T. Bruhns, A. Meyers, J. Elast. **47**, 51 (1997). https://doi.org/10.1023/A:1007356925912
9. R.C. Batra, Int. J. Non Linear Mech. **36**, 421 (2001). https://doi.org/10.1016/S0020-7462(00)00057-3
10. R.C. Batra, *Elements of Continuum Mechanics* (AIAA, Reston, 2006)

11. P. Neff, I.D. Ghiba, J. Lankeit, J. Elast. **121**, 143 (2015). https://doi.org/10.1007/s10659-015-9524-7
12. S.N. Korobeynikov, J. Elast. **143**, 147 (2021). https://doi.org/10.1007/s10659-020-09808-2
13. T.J.R. Hughes, in *Theoretical Foundation for Large-scale Computations for Nonlinear Material Behavior*, eds. by S. Nemat-Nasser et al. (Martinus Nijhoff Publishers, Dordrecht, 1984), pp. 29–63
14. S.H. Lo, Int. J. Numer. Methods Eng. **26**, 121 (1988). https://doi.org/10.1002/nme.1620260109
15. M.B. Rubin, *Continuum Mechanics with Eulerian Formulations of Constitutive Equations* (Springer, Cham, 2021)
16. A. Drozdov, Int. J. Solids Struct. **44**(1), 272 (2007). https://doi.org/10.1016/j.ijsolstr.2006.04.022
17. M.M. Rashid, Int. J. Numer. Methods Eng. **36**, 3937 (1993). https://doi.org/10.1002/nme.1620362302
18. K.W. Reed, S.N. Atluri, Comput. Methods Appl. Mech. Eng. **39**, 245 (1983). https://doi.org/10.1016/0045-7825(83)90094-4
19. K.W. Reed, S.N. Atluri, Int. J. Plast. **1**, 63 (1985). https://doi.org/10.1016/0749-6419(85)90014-2
20. M.B. Rubin, O. Papes, J. Mech. Mater. Struct. **6**, 529 (2011). https://doi.org/10.2140/jomms.2011.6.529
21. M.B. Rubin, Finite Elem. Anal. Des. **175**, 103409 (2020). https://doi.org/10.1016/j.finel.2020.103409
22. E.A. de Souza Neto, D. Peric, D.J.R. Owen, *Computational Methods for Plasticity: Theory and Applications* (Wiley, Chichester, 2008)
23. A. Rodriguez-Ferran, P. Pegon, A. Huerta, Int. J. Numer. Methods Eng. **40**, 4363 (1997). https://doi.org/10.1002/(SICI)1097-0207(19971215)40:23<4363::AID-NME263>3.0.CO;2-Z
24. A. Rodriguez-Ferran, A. Huerta, J. Eng. Mech. **124**, 939 (1998). https://doi.org/10.1061/(ASCE)0733-9399(1998)124:9(939)
25. M. Hollenstein, M. Jabareen, M.B. Rubin, Comput. Mech. **52**, 649 (2013). https://doi.org/10.1007/s00466-013-0838-7
26. M. Jabareen, Int. J. Eng. Sci. **96**, 46 (2015). https://doi.org/10.1016/j.ijengsci.2015.07.001
27. A. Meyers, H. Xiao, O. Bruhns, Comput. Struct. **84**, 1134 (2006). https://doi.org/10.1016/j.compstruc.2006.01.012
28. L. Gambirasio, G. Chiantoni, E. Rizzi, Arch. Computat. Methods Eng. **23**, 39 (2016). https://doi.org/10.1007/s11831-014-9130-z
29. W. Ji, A.M. Waas, Z.P. Bažant, J. Appl. Mech. **80**, 041024 (2013). https://doi.org/10.1115/1.4007828
30. N. Nguyen, A. Waas, Z. Angew. Math. Phys. **67**, 35 (2016). https://doi.org/10.1007/s00033-016-0623-5
31. O.T. Bruhns, Z. Angew. Math. Mech. **94**, 187 (2014). https://doi.org/10.1002/zamm.201300243
32. O.T. Bruhns, Acta. Mech. Sin. **36**, 472 (2020). https://doi.org/10.1007/s10409-020-00926-7

Chapter 7
Concluding Remarks

Abstract In this chapter we summarize and comment on all the research presented in this book.

We have shown that the weak and strong incrementally objective algorithms known in the literature can be obtained as special cases from the generalized midpoint algorithms developed by Simo and Hughes [1] (see Sects. 4.2.3 and 4.3). To provide a theoretical basis for this conclusion, we introduced definitions of the incremental objectivity for (absolutely) Eulerian tensors and showed that when using strong incrementally objective algorithms, all previously published expressions for approximate incremental strain tensors are incremental Eulerian or incremental Lagrangian tensors. At the same time, for both (Eulerian and Lagrangian) versions of strong incrementally objective algorithms, incremental rotation tensors are incremental Eulerian–Lagrangian objective tensors (see Sect. 4.3).

We have also developed new algorithms for integrating CRs for hypoelastic material models, which we called absolutely objective algorithms. Like strong incrementally objective algorithms, these algorithms exactly reproduce stresses in the case of additional superimposed rotations on deformed configurations of bodies. These reproductions of stresses were made possible by ensuring the absolute (non-incremental) objectivity property for all approximate incremental quantities. We have considered both the Lagrangian and Eulerian versions of absolutely objective algorithms and found that the Lagrangian version (theoretically equivalent to the Eulerian version) requires fewer arithmetic operations to determine the Kirchhoff stress tensor compared to the Eulerian version. The Lagrangian version of absolutely objective algorithms was used for computer simulations of simple extension and simple shear (see Sect. 6.5).

Our computer simulations of simple extension and simple shear have shown that strong incrementally objective algorithms are preferred to absolutely objective algorithms for integrating CRs for Hooke-like hypoelasticity based on the Zaremba–Jaumann and Hill stress rates. At the same time, absolutely objective algorithm are preferred to strong incrementally objective algorithms for integrating CRs for Hooke-like hypoelasticity based on the Green–Naghdi and logarithmic stress rates. We attribute this to the fact that the spin tensor associated with the Zaremba–Jaumann and Hill stress rates is the vorticity tensor $\mathbf{w}$, which does not depend on the choice of

© The Author(s), under exclusive license to Springer Nature Switzerland AG 2023
S. Korobeynikov and A. Larichkin, *Objective Algorithms for Integrating Hypoelastic Constitutive Relations Based on Corotational Stress Rates*, SpringerBriefs in Continuum Mechanics, https://doi.org/10.1007/978-3-031-29632-1_7

the reference configuration of the deformable body, and the spin tensors associated with the Green–Naghdi and logarithmic stress rates are the skew-symmetric tensors ω^R and $\omega^{\log}$, which depend on the choice of this configuration (see Chap. 6).

Another novelty of these studies is our recommendation to use the incremental Mooney strain tensors as approximate incremental strain tensors since they fairly accurately approximate incremental logarithmic strain tensors but are easier to determine than the latter. This statement has been confirmed by computer simulations of simple extension and simple shear (see Chap. 6).

Reference

1. J.C. Simo, T.J.R. Hughes, *Computational Inelasticity* (Springer, N.Y., 1998)

Appendix A
Flowcharts of Objective Algorithms for Integrating Hypoelastic Constitutive Relations

For ease of reference, each of the objective algorithms for integrating hypoelastic constitutive relations considered in this book is assigned its own unique ID, which is given in the form ID-Alg in Table A.1. In the same Table A.1, we also provide a reference to the Section of this book in which this algorithm is described. An overview of expressions for incremental kinematic tensors and stress updates for the objective algorithms considered in this book is represented in Table A.2.

Table A.1 Identifier ID-Alg of the objective algorithms for integrating hypoelastic constitutive relations considered in this book

ID-Alg	Algorithm	Section of the book with a description of the algorithm
1	Weak incrementally objective (mid-point), the H–W version	(4.2.1)
2	Weak incrementally objective (mid-point), the R–A version	(4.2.1), (4.2.2)
3	Strong incrementally objective, incrementally Lagrangian version	(4.3.2)–(4.3.4)
4	Strong incrementally objective, incrementally Eulerian version	(4.3.2)–(4.3.4)
5	Absolutely Lagrangian-objective	(5.1)
6	Absolutely Eulerian-objective	(5.2)

© The Author(s), under exclusive license to Springer Nature Switzerland AG 2023
S. Korobeynikov and A. Larichkin, *Objective Algorithms for Integrating Hypoelastic Constitutive Relations Based on Corotational Stress Rates*, SpringerBriefs in Continuum Mechanics, https://doi.org/10.1007/978-3-031-29632-1

Table A.2 Overview of expressions for incremental kinematic tensors and stress updates for the objective algorithms

ID-Alg	Incremental strain [reference]	Incremental rotation [reference]	Stress update [reference]
1	$\Delta_t \mathbf{d}^m$ [Eq. (4.21)]	$^{\Delta t}\boldsymbol{\Psi}$ [Eq. (4.23)]	$^{t+\Delta t}\boldsymbol{\tau}$ [Eq. (4.24)]
2	$\Delta_t \mathbf{d}^m$ [Eq. (4.21)]	$^{\Delta t}\boldsymbol{\Psi}$ [Eqs. (4.25), (4.26)]	$^{t+\Delta t}\boldsymbol{\tau}$ [Eq. (4.24)]
3	$\Delta_t \tilde{\bar{\mathbf{d}}}^a$ [Table 4.1][b]	$^{\Delta t}\tilde{\boldsymbol{\Psi}}_\omega$ [Eq. (4.62)]	$^{t+\Delta t}\boldsymbol{\tau}$ [Eq. (4.67)]
4	$\Delta_t \mathbf{d}^a$ [Table 4.1][b]	$^{\Delta t}\boldsymbol{\Psi}_\omega$ [Eq. (4.60)]	$^{t+\Delta t}\boldsymbol{\tau}$ [Eq. (4.66)]
5	$\Delta_t \mathbf{D}^a$ [Table 5.1][b]	$^{\Delta t}\boldsymbol{\Psi}_\Omega$ [Eq. (5.12)+ Rodrigues's formula]	$^{t+\Delta t}\bar{\boldsymbol{\tau}}$ [Eq. (5.13)]
6	$\Delta_t \tilde{\bar{\mathbf{d}}}^a$ [Table 5.1][b]	$^{\Delta t}\boldsymbol{\Psi}_\omega$ [Eq. (5.26)]	$^{t+\Delta t}\boldsymbol{\tau}$ [Eq. (5.27)]

[b]One can use any corresponding expressions for the tensors $\Delta_t \tilde{\bar{\mathbf{d}}}^a$, $\Delta_t \mathbf{d}^a$, $\Delta_t \mathbf{D}^a$, and $\Delta_t \tilde{\bar{\mathbf{d}}}^a$ from Tables 4.1 and 5.1, but we recommend to explore for these tensors expressions for $\Delta_t \bar{\mathbf{d}}_4$, $\Delta_t \mathbf{d}_4$, $\Delta_t \mathbf{D}_4$, and $\Delta_t \tilde{\bar{\mathbf{d}}}_4$

Flowcharts of the objective algorithms for integrating hypoelastic constitutive relations considered in this paper are presented below. The main flowchart of the objective algorithms is shown in Fig. A.1. Flowchart of weak incrementally objective (mid-point) algorithms with ID-Alg = 1 (H–W version) and 2 (R–A version) is given in Fig. A.2. Flowcharts of strong incrementally Lagrangian- and Eulerian-objective algorithms with ID-Alg = 3 and 4 are given in Figs. A.3 and A.4, respectively. Flowcharts of absolutely Lagrangian- and Eulerian-objective algorithms with ID-Alg = 5 and 6 are given in Figs. A.5 and A.6, respectively.

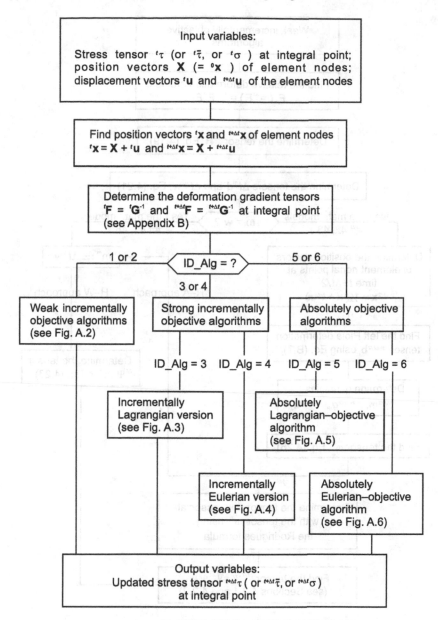

Fig. A.1 Main flowchart of objective algorithms (the ID-Alg corresponds to the ID number of algorithm, see Table A.1)

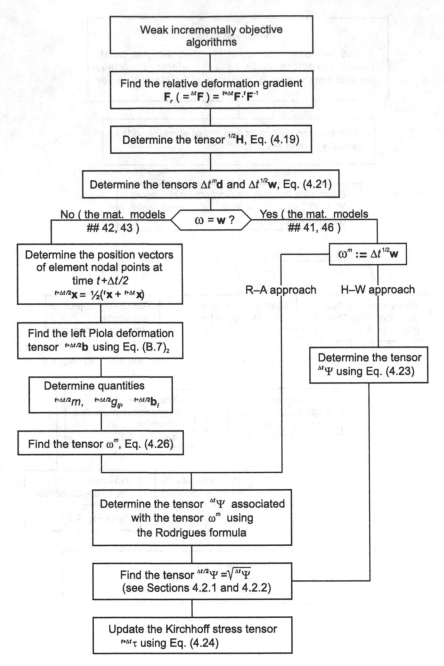

Fig. A.2 Flowchart of weak incrementally objective algorithms with ID-Alg = 1 (H–W version) and 2 (R–A version)

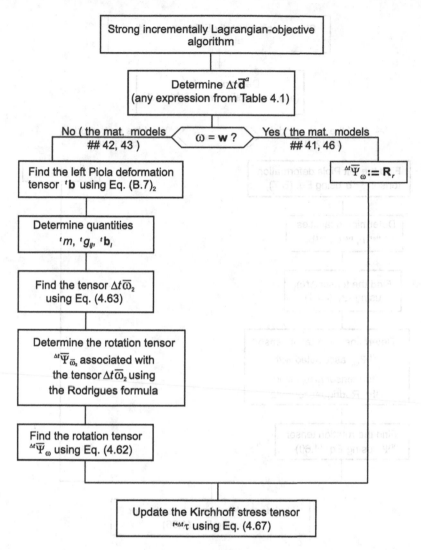

Fig. A.3 Flowchart of strong incrementally Lagrangian-objective algorithm with ID-Alg = 3

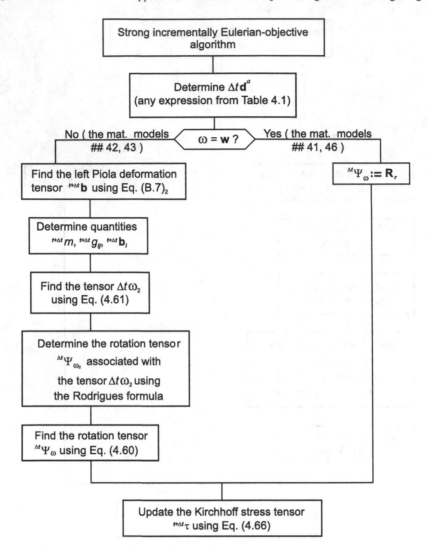

Fig. A.4 Flowchart of strong incrementally Eulerian-objective algorithm with ID-Alg = 4

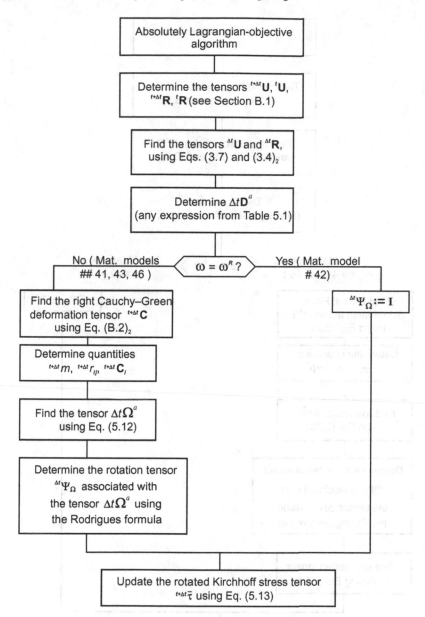

Fig. A.5 Flowchart of absolutely Lagrangian-objective algorithm with ID-Alg = 5

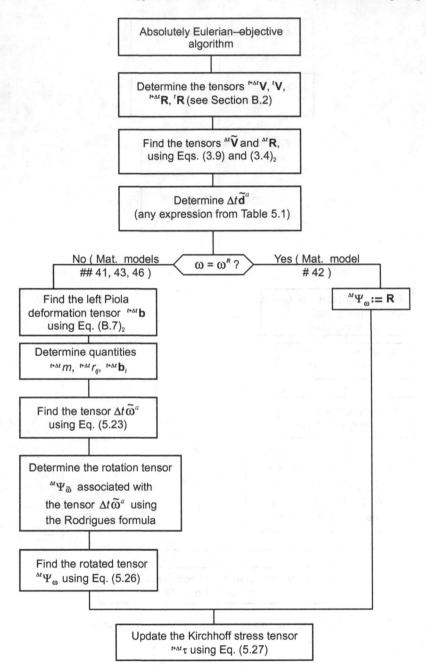

Fig. A.6 Flowchart of absolutely Eulerian-objective algorithm with ID-Alg = 6

Appendix B
Algorithms for Determining the Polar Decomposition

As noted in Remark 2.1, for solid mechanics equations in Lagrangian variables, it is convenient to use the deformation gradient tensor $\mathbf{F}$, and for solid mechanics equations in Eulerian variables, the inverse deformation gradient tensor $\mathbf{G} = \mathbf{F}^{-1}$. Algorithms for determining the polar decomposition in Lagrangian variables using the tensor $\mathbf{F}$ are described in Sect. B.1, and algorithms for determining the polar decomposition in Eulerian variables using the tensor $\mathbf{G}$ in Sect. B.2.

B.1 Algorithms for Determining the Polar Decomposition in Lagrangian Variables

The first algorithm presented in this section is well known. Here we reproduce this algorithm for the sake of self-sufficiency of the matter. In FE analysis, only the position vectors of the nodal points in the reference configuration $\mathbf{X}$ and the displacement vectors $\mathbf{u}$ of the nodal points for each element are usually known. Using FE approximations of the displacement vector, it is relatively easy to find the displacement gradient tensor in Lagrangian variables at some Gauss integration point of the element in question

$$\mathbf{H} \equiv \frac{\partial \mathbf{u}}{\partial \mathbf{X}} \left(= \frac{\partial u_i}{\partial X_j} \mathbf{k}_i \otimes \mathbf{k}_j \right). \tag{B.1}$$

Hereinafter, the quantities $\mathbf{k}_i$ and $\mathbf{k}_j$ ($i, j = 1, \dots 3$) are orthonormal basis vectors of the Cartesian coordinate system. The deformation gradient tensor $\mathbf{F}$ and the right Cauchy–Green deformation tensor $\mathbf{C}$ are found from the equalities

$$\mathbf{F} \left(= \frac{\partial \mathbf{x}}{\partial \mathbf{X}} \right) = \mathbf{I} + \mathbf{H}, \quad \mathbf{C} = \mathbf{F}^T \cdot \mathbf{F} = \mathbf{U}^2. \tag{B.2}$$

© The Author(s), under exclusive license to Springer Nature Switzerland AG 2023
S. Korobeynikov and A. Larichkin, *Objective Algorithms for Integrating Hypoelastic Constitutive Relations Based on Corotational Stress Rates*, SpringerBriefs in Continuum Mechanics, https://doi.org/10.1007/978-3-031-29632-1

Determining the eigenindex m, eigenvalues μ_i, and corresponding eigenprojections C_i ($i, j = 1, \dots m$) of the tensor C, one can obtain the spectral representation $(2.3)_1$ of this tensor. Then for the tensor U, the spectral representation $(2.2)_1$ holds if the equalities $(2.4)_{1,2}$ are satisfied. The tensor R is obtained from the equality $(2.1)_1$

$$R = F \cdot U^{-1}, \quad U^{-1} = \sum_{i=1}^{m} \lambda_i^{-1} C_i \ (\lambda_i = \sqrt{\mu_i}). \tag{B.3}$$

An alternative algorithm for determining the tensors U and R is based on finding the principal invariants of the tensor C and applying the Caley–Hamilton theorem. This algorithm does not require determining principal eigenvalues, but in the case of 3D analysis, it seems to be too cumbersome compared to the above algorithm based on the spectral representation of the tensor C. However, in the case of 2D analysis, the latter algorithm has advantages over the former one since it allows obtaining compact expressions for determining the tensors U and R. We present explicit expressions for the tensor U^{-1}

$$U^{-1} = -I_U[II_U(II_U + I_C) + II_C]^{-1}[C - (II_U + I_C)I], \tag{B.4}$$

where

$$I_C \equiv \operatorname{tr} C, \quad II_C \equiv \det C, \quad I_U = \sqrt{I_C + 2\sqrt{II_C}}, \quad II_U = \sqrt{II_C}. \tag{B.5}$$

The tensor R is still determined from the expression (B.3).

B.2   Algorithms for Determining the Polar Decomposition in Eulerian Variables

Assuming that for each finite element, the position vectors of the nodal points of the element in the reference configuration X and the displacement vectors of the nodal points u of the element are known, we find the position vectors of the nodal points of the element in the current configuration $x = X + u$. Using FE approximations of the displacement vector, it is relatively easy to find the displacement gradient tensor in Eulerian variables at some Gauss integration point of the element in question (see, e.g., [1])

$$K \equiv \frac{\partial u}{\partial x} \left(= \frac{\partial u_i}{\partial x_j} k_i \otimes k_j \right). \tag{B.6}$$

The inverse deformation gradient tensor $G \ (= F^{-1})$ and the left Piola deformation tensor b are found from the equalities

$$G\left(=\frac{\partial X}{\partial x}\right) = I - K, \quad b = G^T \cdot G(= F^{-T} \cdot F^{-1}) = V^{-2}. \tag{B.7}$$

Determining the eigenindex m, eigenvalues κ_i, and corresponding eigenprojections b_i $(i, j = 1, \ldots m)$ of the tensor b, one can obtain the spectral representation $(2.3)_2$ of this tensor. Then for the tensor V, the spectral representation $(2.2)_2$ holds if the equalities $(2.4)_{3,4}$ are satisfied. The tensor R is found from the equality $(2.1)_1$

$$R = V \cdot G^T (= V \cdot F^{-T}). \tag{B.8}$$

An alternative algorithm for determining the tensors V and R is derived from the similar algorithm for determining the tensors U and R given in Appendix B.1. In particular, for 2D analysis, the tensor V can be found from the expression $(Y \equiv V^{-1})$

$$V = Y^{-1} = -I_Y[II_Y(II_Y + I_b) + II_b]^{-1}[b - (II_Y + I_b)I], \tag{B.9}$$

where

$$I_b \equiv \operatorname{tr} b, \quad II_b \equiv \det b, \quad I_Y = \sqrt{I_b + 2\sqrt{II_b}}, \quad II_Y = \sqrt{II_b}. \tag{B.10}$$

The tensor R is still determined from the expression (B.8).

Appendix C
The Rodrigues Formula for Determining an Incremental Rotation Tensor

The Rodrigues formula is used to obtain an approximate solution of the Cauchy problem in the time interval $[t, t + \Delta t]$

$$\dot{\Psi} = \omega \cdot \Psi, \quad \Psi = I \text{ at time } t, \tag{C.1}$$

where $\omega \in \mathcal{T}^2_{\text{skew}}$ and $\Psi \in \mathcal{T}^{2+}_{\text{orth}}$ are the spin tensor and its associated rotation tensor, respectively. Let ω^* be the known value of the tensor ω at some time $t + \alpha \Delta t$ in the time interval $[t, t + \Delta t]$ ($\alpha \in [0, 1]$); i.e., the following equality holds:

$$\omega^* = {}^{t+\alpha\Delta t}\omega. \tag{C.2}$$

We assume that in the time interval $[t, t + \Delta t]$, the tensor ω is constant and equal to its value at the time $t + \alpha \Delta t$; i.e., the following equality holds:

$$\omega(t) = \omega^* \text{ at } t \in [t, t + \Delta t]. \tag{C.3}$$

We approximate the solution of the nonlinear Cauchy problem (C.1) by the solution of the linear one

$$\dot{\Psi} = \omega^* \cdot \Psi, \quad \Psi = I \text{ at } t = t_0. \tag{C.4}$$

The exact solution of the problem (C.4) has the form

$$^{\Delta t}\Psi = \exp(\Delta t\, \omega^*) \text{ at time } t + \Delta t. \tag{C.5}$$

This solution can be represented by the *Rodrigues formula*

$$^{\Delta t}\Psi = I + (\sin \omega \Delta t)\overline{\omega} + (1 - \cos \omega \Delta t)\overline{\omega}^2. \tag{C.6}$$

© The Author(s), under exclusive license to Springer Nature Switzerland AG 2023 103
S. Korobeynikov and A. Larichkin, *Objective Algorithms for Integrating Hypoelastic Constitutive Relations Based on Corotational Stress Rates*, SpringerBriefs in Continuum Mechanics, https://doi.org/10.1007/978-3-031-29632-1

where[1]

$$\overline{\omega} \equiv \omega^*/\omega, \quad \omega^2 \equiv \frac{1}{2}\omega^* : \omega^* = (\omega_{12}^*)^2 + (\omega_{13}^*)^2 + (\omega_{23}^*)^2. \tag{C.7}$$

Hereinafter, ω_{12}^*, ω_{13}^*, and ω_{23}^* are generally nonzero Cartesian components of the tensor ω^*.

In particular, for 2D analysis, we set $\omega_{13}^* = \omega_{23}^* = 0$; then from $(C.7)_2$ we obtain

$$\omega = |\omega_{12}^*| = |\omega_{21}^*|. \tag{C.8}$$

and from (C.6) we obtain the following simple expression for the incremental rotation tensor:

$$^{\Delta t}\boldsymbol{\Psi} = \begin{bmatrix} \cos(\omega_{12}^*\Delta t) & \sin(\omega_{12}^*\Delta t) & 0 \\ -\sin(\omega_{12}^*\Delta t) & \cos(\omega_{12}^*\Delta t) & 0 \\ 0 & 0 & 1 \end{bmatrix} = \begin{bmatrix} \cos(\omega_{21}^*\Delta t) & -\sin(\omega_{21}^*\Delta t) & 0 \\ \sin(\omega_{21}^*\Delta t) & \cos(\omega_{21}^*\Delta t) & 0 \\ 0 & 0 & 1 \end{bmatrix}. \tag{C.9}$$

The tensor $^{\Delta t/2}\boldsymbol{\Psi}$ can be determined by replacing Δt with $\Delta t/2$ in the above expressions.

Reference

1. K.J. Bathe, *Finite Element Procedures* (Prentice Hall, Upper Saddle River, New Jersey, 1996)

---

[1] Herein, the operation ":" denotes the double inner product (double contraction) of tensors $\mathbf{A}$, $\mathbf{H} \in \mathcal{T}^2$: $\mathbf{A} : \mathbf{H} \equiv \mathrm{tr}(\mathbf{A} \cdot \mathbf{H}^T)$.

Index

Printed in the United States
by Baker & Taylor Publisher Services